AVID

READER

PRESS

CIVILIZED TO DEATH

THE PRICE OF PROGRESS

CHRISTOPHER RYAN

AVID READER PRESS

NEW YORK LONDON TORONTO SYDNEY NEW DELHI

Avid Reader Press
An Imprint of Simon & Schuster, Inc.
1230 Avenue of the Americas
New York, NY 10020

First Avid Reader Press trade paperback edition August 2020

AVID READER PRESS and colophon are trademarks
of Simon & Schuster, Inc.

For information about special discounts for bulk purchases,
please contact Simon & Schuster Special Sales at 1-866-506-1949
or business@simonandschuster.com.

The Simon & Schuster Speakers Bureau can bring authors to
your live event. For more information or to book an event, contact
the Simon & Schuster Speakers Bureau at 1-866-248-3049
or visit our website at www.simonspeakers.com.

Interior design by Kyle Kabel

Manufactured in Italy

5   7   9   10   8   6

Library of Congress Cataloging-in-Publication Data:
https://lccn.loc.gov/2018278079

ISBN 978-1-4516-5910-8
ISBN 978-1-4516-5911-5 (pbk)
ISBN 978-1-4516-5912-2 (ebook)

To Frank and Julie

*And to those who have been, are being,
and will be civilized to death*

The friendly and flowing savage, who is he? Is he waiting for civilization, or past it and mastering it?

—Walt Whitman

Contents

CONTENTS

Part II: APOCALYPSE ALWAYS (THE NPP IN THE PRESENT)

Part III: REFLECTIONS IN AN ANCIENT MIRROR (BEING HUMAN)

CONTENTS

Part IV: A PREHISTORIC
PATH INTO THE FUTURE

Introduction: Know Thy Species

Call me ungrateful. I've got silver fillings in my teeth, artisanal beer in my fridge, and a world of music in my pocket. I drive a Japanese car with cruise control, power steering, and air bags poised to cushion me in an explosive embrace should I drift off. I wear German glasses that darken in California sunlight, and I'm writing these words on a computer that's thinner and lighter than the book they'll eventually be printed in. I enjoy the company of friends I'd have lost if they hadn't been saved by emergency surgery, and, for the last seventeen years of his life, my father's blood was filtered through the liver of a man named Chuck Zoerner, who died in 2002. I have every reason to appreciate the many wonders of civilization.

And yet.

When the English author G. K. Chesterton first visited America, in 1921, his hosts took him to see Times Square at night. Chesterton stood staring in silence for several increasingly awkward moments. When someone finally asked him for his thoughts, Chesterton replied: "I was thinking how beautiful this would be if I couldn't read."

Like Chesterton, we can read the signs, and they're not good. The insistent, flashing ads are steadily losing their power to distract us from what many know and most suspect: We're approaching the end of the road. Belief in progress—the promise and premise of civilization—is melting away like a glacier.

But what about antibiotics and airplanes, women's rights, gay marriage? True enough. But upon closer inspection, many of the supposed gifts of civilization turn out to be little more than partial compensation for what we've already paid, or they cause as much trouble as they claim to solve.

Most of the infectious diseases vaccines protect us from, for example, were never a problem until humans began living with domesticated animals from which pathogens jumped over to our species. Influenza, chicken pox, tuberculosis, cholera, heart disease, depression, malaria, tooth decay, most types of cancer, and just about every other major ailment responsible for causing massive suffering to our species derive their lethality from some aspect of civilization: domesticated animals, densely populated towns and cities, open sewers, food contaminated with pesticides, disruptions to our microbiome, and so on.

Within just a few years of unlocking the miracle of flight, pilots were flying with one hand while tossing bombs on civilians with the other. And only in the most progressive modern societies are LGBTQ people and women regaining the acceptance and respect they typically received in most foraging societies.* Reports of progress have a tendency to be wildly overstated and uncritically accepted, while anyone who questions the benefits of civilization is liable to be dismissed as cynical, utopian, or some hybrid of both.

"An era can be considered over," said Arthur Miller, "when its basic illusions have been exhausted." Progress, surely the basic illusion of our era, is spent. Dystopian scenarios loom ever larger as fisheries collapse, CO_2 levels rise, and clouds of radioactive

---

* I'll use "forager," "hunter-gatherer," and "uncivilized" interchangeably, to avoid repetition. In every case, unless otherwise noted, I'm referring to what anthropologists call "immediate return hunter-gatherers," people who do not typically accumulate food, but eat what they find as they find it.

steam billow from "fail-safe" nuclear plants. Oil gushes into oceans, mutating pathogens overwhelm the last effective antibiotics, and the living dead stumble through our collective unconscious. Each successive year is the hottest on record, and the next undeclared war ignites from the embers of the previous while political parties nominate charlatans who can't agree on what's happening, much less what to do about it. Despite the marvels of our age—or maybe partly because of them—these are deeply troubled times.

It's common to wonder what sage advice an emissary from the future might bring back to help us choose the best path forward. But consider the opposite scenario. How would a time traveler from the prehistoric past assess the state and trajectory of the modern world? She would no doubt be impressed by much of what she encountered here, but once her amazement at mobile phones, air travel, and self-driving cars subsided, what would she make of the substance and meaning of our lives? Would she be more awed by our doodads or dismayed by what we've left behind in our rush toward an increasingly precarious future?

This question isn't as hypothetical as it seems. Missionaries, explorers, adventurers, and anthropologists have been consistently confused and disappointed by indigenous people's rejection of the comforts and constraints of civilization. "Why should I learn to farm," asked a !Kung San man, "when there are so many mongongo nuts in the world?" In a letter to a friend, Benjamin Franklin noted how little interest Indians had in joining civilization: "They have never shown any inclination to change their manner of life for ours. When an Indian child has been brought up among us, taught our language and habituated to our customs, yet if he goes to see his relations and make one Indian ramble with them, there is no persuading him ever to return." And when white children got a taste of Indian life (generally due to having been kidnapped), they also preferred it, according to Franklin.

After their rescue, "in a short time they become disgusted with our manner of life, and the care and pains that are necessary to support it, and take the first good opportunity of escaping again into the woods."

Charles Darwin saw firsthand how difficult it was to sell native people on civilization. Passing through Tierra del Fuego on the *Beagle*, he was amazed by what seemed to him to be the squalor and degradation of the people living at the cold and stormy southernmost tip of the Americas. In a letter to a friend, Darwin wrote: "I have seen nothing, which more completely astonished me, than the first sight of a Savage; It was a naked Fuegian his long hair blowing about, his face besmeared with paint." In his journal, Darwin wrote, "I believe if the world was searched, no lower grade of man could be found."

On an earlier trip, the *Beagle*'s captain, Robert FitzRoy, had abducted three Fuegians, two children—whom the British called Fuegia Basket and Jemmy Button—and a young man they called York Minister. The kidnapping was justified, FitzRoy felt, because "the ultimate benefits arising from their acquaintance with our habits and language, would make up for the temporary separation from their own country." FitzRoy had taken them back to England, where they spent over a year being indoctrinated into civilized life—even meeting King William IV and Queen Adelaide during their stay. Now familiar with the obvious superiority of European society, they were on the *Beagle* with Darwin, headed back to their own people in Tierra del Fuego so they could preach the good word concerning the proper, civilized approach to life.

But when the *Beagle* returned to Woolya Cove, near what is now called Mount Darwin, just a year after dropping them off, Jemmy, York, and Fuegia were nowhere to be found. The huts and gardens the British sailors had built for the three Fuegians were deserted and overgrown. Eventually, Jemmy was located and joined Darwin and FitzRoy for dinner on the ship, where he confirmed that the Fuegians had abandoned their civilized ways. Overcome with sadness, Darwin wrote that he'd never seen "so complete & grievous a change" and that "it was painful to behold him." (Darwin noted that Jemmy hadn't forgotten how to use a knife and fork properly.) When Captain FitzRoy offered transport back to England, Jemmy declined, saying he had "not the least wish to return to England" as he was happy and contented with "plenty fruits," "plenty fish," and "plenty birdies."*

* * *

Carl Jung lamented our drift from the past and the "uprootedness" that led people to live "more in the future and its chimerical promises of a golden age than in the present, with which our whole evolutionary background has not yet caught up." Writing in his memoir, *Memories, Dreams, Reflections*, Jung couldn't have been clearer in lamenting our species' drift into future-fantasy: "We rush impetuously into novelty, driven by a mounting sense of insufficiency, dissatisfaction, and restlessness. We no longer live on what we have, but on promises, no longer in the light of the present day, but in the darkness of the future, which, we expect, will at last bring a proper sunrise. We refuse to recognize that everything better is purchased at the price of something worse."

In a 1928 essay called "Economic Possibilities for Our Grand-

---

* Twenty-five years later, in late 1859, within days of the publication of *On the Origin of Species by means of Natural Selection or the Preservation of Favored Races in the Struggle for Life*, Jemmy Button led an attack on a party of Christian missionaries in Tierra del Fuego, killing eight of them. And FitzRoy? After delivering young Charles Darwin and his revolutionary ideas back to England, Captain FitzRoy invented the science of weather forecasting and brought about a revolution in meteorology. But for all his scientific accomplishments, FitzRoy remained a deeply pious man, and the publication of *Origin of Species* mortified him.

children," the famous economist John Maynard Keynes imagined the world a century into the future. Things would be so good, he predicted, that no one would need to worry about making money. The principal problem people would face would be figuring out what to do with their overwhelming amount of free time: "For the first time since his creation man will be faced with his real, his permanent problem," Keynes wrote, "how to use his freedom from pressing economic cares, how to occupy the leisure, which science and compound interest will have won."

Well, here we are in that much anticipated future, and the average American is as frazzled and desperate as ever, working as many hours today as he or she did in 1970 and lucky to get a couple of weeks off per year. It's technically true that measures of global wealth are up in the past few decades, but, at least in Europe and the United States, almost all the surplus wealth has gone to those who need it least, leaving the rest further behind than ever.

And not even the luckiest among us are really all that comfortable. Forty-four percent of Americans earning between $40,000 and $100,000 per year told researchers that they couldn't come up with $400 in an emergency, and 27 percent of those making more than $100,000 said the same. Globally, gross domestic product (GDP) increased 271 percent between 1990 and 2014, yet the number of people living on less than five dollars a day rose 10 percent in the same period, and the number of people going hungry increased by 9 percent.

Ah, the glorious, leisurely future—always just around the corner. Think I'm being too harsh? Evolutionary biologist Stephen Jay Gould called the very notion of progress "a noxious, culturally embedded, untestable, nonoperational, intractable idea that must be replaced if we wish to understand the patterns of history." While a bit more diplomatic, Jared Diamond isn't convinced by the pro-progress propaganda, either, suggesting that words such as "civilization" and the phrase "the rise of civilization" falsely imply "that civilization is good, tribal hunter-gatherers are miserable, and history for the past 13,000 years has involved progress toward greater human happiness." But Diamond doesn't buy it, writing, "I do not assume that industrialized states are 'better' than hunter-gatherer tribes, or that the abandonment of the hunter-gatherer lifestyle for iron-based statehood represents 'progress,' or that it has led to an increase in human happiness."

But I hear the progress lovers, the true believers in the self-evident notion that we're fulfilling our destiny as the planet's chosen species by progressing toward some asymptotic goal that grows ever closer—even if it never quite arrives. I don't dispute the reality of progress in certain contexts, but I have my doubts about how to conceptualize and measure it. We tend to confuse progress with adaptation, for example. Adaptation—and, by extension, evolution—doesn't presuppose that a species is getting "better" as it evolves, merely that it is growing more suited to its environment. The "fittest" may survive and reproduce, but "fitness" is a concept that exists only within a specific ecological context, having no absolute, noncontextual meaning or value. Male Egyptian vultures, for example, smear shit all over their faces—presumably to demonstrate their immunological prowess to females. This particular fitness display is probably not so effective in other species.

It often seems to me that we are progressing either toward a modern manifestation of our distant past or toward a precipice. Our desperate peregrinations are in search of a place much like the home we left when we walked out of the garden and started to farm. Our most urgent dreams may simply reflect the world as it was before we fell asleep.

Perhaps we're approaching the so-called singularity, when our comfort-atrophied bodies melt into the screens we spend so much of our lives staring into. Or perhaps the colonization of other planets will allow our descendants to live in distant domes sponsored by Apple, Tesla, and Caesars Palace. If you, like Keynes, were hoping for an egalitarian world of shared plenitude and lots of free time to enjoy the company of those you love, consider that our ancestors occupied a world very much like that until the advent of agriculture and what came to be called "civilization" sprouted about ten thousand years ago, and we've been progressing away from it ever since.

When you're going in the wrong direction, progress is the last thing you need. The "progress" that defines our age often seems closer to the progression of a disease than to its cure. Civilization often seems to be picking up speed in the dizzying way things do when they're circling the drain. Could it be that the fiercely held belief in progress is a sort of painkiller—a faith-in-the-future antidote to a present too terrifying to contemplate?

I know, there's always been some lunatic warning that the end was nigh and he's always said, "This time it's different!" But seriously, this time it's different. Headlines like "We're Doomed. Now What?" loom from the pages of major newspapers. The planetary climate is shifting like cargo on a sinking ship. The UN high commissioner for refugees reports that at the end of 2015, the number of people forcibly displaced by war, conflict, and persecution had risen to a staggering 65.3 million, up from 37.5 million in 2004. Flocks of birds are falling dead from the sky, the buzzing of bees is fading, butterfly migrations have stopped, and vital ocean currents are slowing. Species are going extinct at a rate not seen since the dinosaurs vanished 65 million years ago. Texas-sized masses of swirling plastic soup suffocate acidifying oceans while freshwater aquifers are pumped dry as a bone. Ice caps melt down as clouds of methane bubble up from the depths, accelerating the cycle of global destruction. Governments look away while Wall Street tears the last bits of wealth from the carcass of the middle class and energy companies frack the earth, pumping secret poisons into aquifers we all depend on but don't know how to protect. Little wonder that depression is the leading cause of disability in the world, and is growing quickly.

The state of things is shocking and worrisome, but shouldn't surprise us. Every civilization that's ever existed has collapsed into chaos and confusion. Why presume that ours will break the pattern? But there *is* a difference: While Rome, Sumer, the Mayans, ancient Egypt, Easter Island, and the others ended in regional collapses, the civilization imploding around us now is global. As Canadian historian Ronald Wright put it, "Each time history repeats itself, the price goes up."

Maybe you think that the end of the world is beside the point. Perhaps the sublime beauty of Beethoven's late quartets, photos of Earth taken from space, or knowledge of the structure of DNA are worth any price to you—even the otherworldly price we and the other creatures on this planet are paying. Maybe your life, or the life of someone you love, was saved by technological medicine—which makes it both confusing and distasteful for you to be anything less than a full-throated fan of progress. Maybe you have faith that self-organizing coalitions of smart, decent people will find a way to make corrective memes go viral—rapidly infecting our species, just in time, with some common fucking sense.

Whether the wonders of our age are worth their exorbitant cost is a question each of us must ultimately answer for ourselves. But before we can begin to answer such a crucial question, we must first cut through the veil of pro-progress propaganda to which we've been subjected for centuries in order to do two things: get a fuller conception of civilization that includes its costs and victims, and think hard about how much meaning and fulfillment "modern wonders" actually bring to our lives. If everything's so amazing, why are so many of us so profoundly unhappy?

The widespread belief that noncivilized human life was and is a desperate struggle for survival resonates with the haughty dismissal of uncivilized "savages" common to previous centuries. But beyond its inaccuracy and racist undertones, this view has disastrous consequences in the present. Life-and-death medical decisions are misinformed by false assumptions about the capacities of the human body, relationships fall short of unrealistic expectations, legal systems based on inaccurate notions of human nature generate the very suffering they're meant to avert, educational institutions smother the innate curiosity of students, and so on. Indeed, nearly every aspect of our lives (and our deaths) is distorted by a misinformed sense of what kind of animal *Homo sapiens* really is.

Dr. Jonas Salk, famous for having invented the polio vaccine, put it memorably: "It is necessary now not only to 'know thyself,' but also to 'know thy species' and to understand the 'wisdom' of nature, and especially living nature, if we are to understand and help man develop his own wisdom in a way that will lead to life of such quality as to make living a desirable and fulfilling experience."

But how many of us know our species well enough to know ourselves? For centuries, we've been misinformed about what kind of creature we were, are, and can be. The resulting confusion undermines our attempts to live "desirable and fulfilling" lives. Lies can be repeated so frequently that they become indistinguishable from the voices in our heads: *Civilization is humankind's greatest accomplishment. Progress is undeniable. You're lucky to be alive here and now. Any doubt, despair, or disappointment you feel is your own fault. Get over it. Walk it off. Take a pill and stop complaining.*

To be clear, I harbor no illusions about "noble savages" or "getting back to the garden." To the extent the savages are or ever were noble, we'll see that it's because their societies flourished by promoting generosity, honesty, and mutual respect—values, not coincidentally, still cherished by most modern humans at a gut level. There were concrete, survival-based reasons for our highly interdependent hunter-gatherer ancestors to honor these values and personality characteristics—and for evolution to promulgate them through sexual selection because women found them to be attractive qualities in men. As for Paradise, it's long since been paved over. We've come too far, and there's no going back. Human population levels long ago surpassed the carrying capacity of hunter-gatherer ways that require population densities lower than one person per square mile in most ecosystems. In any case, we're no longer the undomesticated beings our prehistoric ancestors were. We've lost too much of the knowledge and physical conditioning necessary to live comfortably under the stars. If our ancestors were wolves or coyotes, most of us are closer to pugs or poodles.

Years ago, I stumbled upon what might be the saddest zoo in the world, in Bukittinggi, on the Indonesian island of Sumatra. The place was nothing more than a bunch of dismal concrete cages in which a few doomed orangutans languished. I'll never forget the look in their eyes, as they reached out to me from behind rusted iron bars, begging for release, contact, death . . . anything but more of the same. After this intimate look at animals suffering from what I later learned is sometimes called "zoochosis," I didn't go near a zoo again for decades. But eventually, a friend persuaded me to visit the bonobos in San Diego. To call both facilities "zoos" is to highlight the paucity of language. Whatever your opinions concerning animals in captivity, the San Diego Zoo reflects a serious desire to create an artificial world that is as similar to the environments in which each species evolved as possible. The people who designed the enclosures had clearly studied the natural contexts and behavior of the animals destined to live there.

Native habitats were re-created, allowing at least a simulacrum of wildness within the walls.

It's difficult to settle on one element that sets *Homo sapiens sapiens* apart from all other animals. The list of failed candidates is long and includes things like tool use, cultivation of other species for food, nonreproductive sexual behavior, eye contact during sex, female orgasm, organized group conflict, and transmission of knowledge from one generation to the next. Here's my pitch: We are the only species that lives in zoos of our own design. Each day, we create the world we and our descendants are going to inhabit. If we want that world to be more like the San Diego Zoo than the living tombs in Bukittinggi, we'll need a clearer understanding of what human life was like before our ancestors first woke up in cages. We'll need to know our species.

Part I

ORIGIN STORIES

———

We have plunged down a cataract of progress which sweeps us on into the future with ever wilder violence the farther it takes us from our roots.

—Carl Jung

This book tells the story of a story—of *the* story. Before civilization, before our ancestors ever blew ochre onto cave walls, before they controlled fire even, they were enthralled by stories. The first human invention is still the most powerful. Who tells the story creates the world.

I misunderstood a key part of the story. I was confused by people who talked about cruelty and ruthlessness and then, with a knowing nod, added, "Well, it's a doggy-dog world." I nodded along with them, but inside I was thinking, "I don't know. A doggy-dog world sounds pretty good to me." Eventually, through tears of laughter, a teacher explained that it is, in fact, a "dog-eat-dog world," after I'd misused the term in a paper.

We tell stories about what happened, but, just as often, the stories we tell determine what happens. Narrative becomes paradigm, because origin stories are as predictive and constraining as they are explanatory. The map showing where we came from delimits where we can go from here. If yours is a story of victimization, you will live out your days a victim. If it's a tale in which your race is superior to all others, evidence of their inferiority will seem plentiful and obvious. To get a clear sense of the future prospects for a relationship, ask the couple how they met. Are they telling a story that arcs toward or away from kindness, mutual respect, and joy? A tale of intractable foes locked in a power struggle isn't going to end "happily ever after."

Here's the story we've all been told about who we are and where we came from:

We are descended from prehistoric ancestors whose lives were a constant struggle against starvation, disease, predators, and each other. Only the strongest, cleverest, most anxious, and most ruthless survived to pass their genes into the future—and even these lucky ones lived only to the age of thirty-five or so. Then, about ten thousand years ago, some forgotten genius invented agriculture, and thus delivered our species from animal desperation into civilized abundance, leisure, sophistication, and plenitude. Despite occasional setbacks, things have been getting better ever since.

In 1651, Thomas Hobbes described human life before the advent of the state as *"solitary, poor, nasty, brutish, and short."* Three and a half centuries later, it remains one of the most famous phrases in the English language, and this Hobbesian vision of our precivilized past persists as the central premise of the story of civilization. This Narrative of Perpetual Progress (NPP) claims to explain the superiority of civilization while taking it as a given. Inasmuch as faith in "progress" is just another way of saying today is ipso facto better than yesterday, the notion of progress is to be taken on faith, and cannot be questioned without inviting the wrath of the true believers. But the NPP poisons our minds, bodies, and relationships as assuredly as it does the air, water, and soil. It justified millennia of slavery and centuries of colonialism.

It generates deep distrust of ourselves and each other, shame and disgust toward our animal bodies, and irrational fear and hostility toward the natural world we are told is out to get us. It insists that we should be grateful for all this progress we've made because, by definition, the here and now is the best time ever to be alive.

The clear implication is that any discontent or despair you may be experiencing *must be* due to some fault of your own—certainly not to the civilization you were born into. You aren't working hard enough, consuming the right products, taking the right supplements, following the right exercise regimen, driving the right car, or drinking enough water.

A recent article in *Scientific American* provides a typical example of these neo-Hobbesian assumptions, warning that "our ancestors were not at one with nature. Nature tried to kill them and starve them out." See that? Nature hates humans! A 2014 book called *Utopia for Realists* begins, "Let's start with a little history lesson: In the past, everything was worse. For roughly 99% of the world's history, 99% of humanity was poor, hungry, dirty, afraid, stupid, sick, and ugly." Ugly, too? A recent article in *Business Insider* begins: "Humanity is always moving forward with innovation after innovation improving global quality of life. . . . Technological advances have boosted productivity by allowing workers to get more done in a day. This helps to increase output and boosts economic growth." Such examples abound, and they all tell the same story: Everything used to be worse than it is now. Thanks to civilization, things for our species have been getting better, and continue to do so.

Note that Hobbes's powerful phrase serves as a condemnation of both the outer conditions of life before the state and of the inner qualities of the brutes themselves. According to this story, our ancestors were horrible, desperate creatures living horrible, desperate lives. This belief that human nature tends toward nastiness, brutality, and suspicion unless countered by the "civilizing" influences of authoritarian institutions is strikingly similar to the idea of original sin—rebranded as science. As with original sin, all human beings are born into a kind of psychological debt, carrying a burden of shame, self-disgust, and suspicion. This pernicious nonsense is self-replicating and self-fulfilling: As a result of having been indoctrinated into this web of toxic beliefs about what sort of creature *Homo sapiens* is, we can find ourselves behaving like the nasty brutes we've been taught to believe we are. To break free of behaviors and beliefs that perpetuate conflict between our inner and outer nature, it's essential to take another look at the Narrative of Perpetual Progress, which overstates the benefits of civilization while ignoring many of its costs and dismissing even respectful doubt as sacrilege.

In this first part of the book, we'll survey salient information about how our ancestors actually lived. In later sections, we'll get into some of the ways the misinformed worldview provided by the NPP can generate trauma, confusion, and suffering in modern lives, and in Part IV, we'll turn our attention to how our species might be starting to tell a new, more accurate origin story that allows for far happier endings than the NPP. If we learn to tell the right story, we may indeed find that our future can be more doggy-dog than dog-eat-dog.

Chapter 1

What We Talk About When We Talk About Prehistory

We've never met, but you and I know each other pretty well. We each have a good sense of what makes the other happy or sad, healthy or ill, aggressive or nurturing. And we've got educated guesses on the parts we don't know for sure: what kinds of foods will make us salivate, which sexual fantasies are most likely to get us hot, what sound patterns soothe us or make us get up and dance, how much and what kind of exercise we need to be fit, the nature of our frustrations over politics and spirituality, and approximately how long we can expect to live. "Show me where you're from," they say, "and I'll tell you who you are." Well, I'm from Africa, at least three hundred thousand years ago. So are you.

You'd be wise to question definitive statements about prehistory—including mine—but there are several windows that offer surprisingly reliable glimpses into the distant past of our species. First-contact accounts and anthropological research have revealed near universalities among recent and present-day hunter-gatherer societies. Because hunter-gatherers in diverse environments behave in strikingly similar ways, most theorists agree that these consistencies are structural—the logical result of how foragers relate to their material environment, a relationship that has remained consistent as far back as science can shine a light.

Some object to this line of reasoning as insulting to contemporary foragers in that it arguably reduces them to "primitives" and misses the indisputable fact that anyone alive today (including contemporary foragers) is every bit as evolved as you and I. This legitimate point doesn't invalidate extrapolating from modern foragers to better understand prehistoric people who interacted with their environments in very similar ways. This line of reasoning is no more outlandish than pointing out commonalities in how baseball players strategized and interacted on the field a hundred years ago and today. Of course there is a lot we cannot know about their motivations and inner lives, but as long as we're sure they were playing with more or less the same rules, we can safely assume quite a bit about their shared approaches to the game.

Furthermore, there is no evidence to suggest the foraging way of life is less sophisticated or satisfying than any other—including our own—and having lasted hundreds of thousands of years, it is certainly more sustainable. I don't share the ubiquitous assumption that twenty-first-century techno sapiens are the pinnacle of biology, or that our species is progressing ever closer to some exalted future state, even further from a past characterized by animalistic misery and desperation. As the title of this book suggests, I am in fact deeply skeptical of such assumptions. Contemporary foragers have been evolving as long as anyone else, but many of the ways they interact with their environments have not changed much for tens of thousands of years. Most of the daily activities of contemporary foragers from the Australian desert to the Arctic Circle have remained remarkably consistent since preagricultural times, including how they hunt, gather, prepare food, build their shelters, make collective decisions, resolve conflict, educate their children, and so on. To conclude that this observation is somehow insulting to foragers requires the assumption that their cultural stability has been detrimental to their quality of life. The evidence, we'll see, does not support such an assumption.

Another window into the past is offered by the human body and its many anatomical and physiological characteristics that reflect the accumulated experiences of our ancestors. For example, if you sit on a toilet, you're doing it wrong. I shit you not. Like the other primates, our bodies are "designed"* to squat when we defecate. The toilet defeats that evolved design, often resulting in hemorrhoids, constipation, and other sorts of unpleasantness.

Evolutionary biologists can read the modern body like a map of human prehistory. The shape, spacing, and hardness of our teeth, the chemicals in our saliva, and the twists and turns of our intestines . . . all tell us a great deal about how and what our ancestors ate. Similarly, from the architecture of our brains to our fascinating genitalia and the arched soles of our feet, our bodies tell stories of the accrued experiences of distant predecessors.

And it's not just our bodies that are etched by the flow of time through our species. Many of our behaviors and biases are reflections of the ancient worlds our ancestors called home. Repeated behaviors become innate tendencies and what we might think of as biological/ psychological "expectations" over millions of sunrises and sunsets, so it's no wonder that pretty much every human being alive would be mesmerized by the dancing flames of a small fire. Virtually every one of our ancestors spent every night of their lives captivated by the same comforting dance. Similarly, there's a remarkable consistency to the social habits of our ancestors—every one of whom was a hunter-gatherer until very recently, in evolutionary time. These innate tendencies often build up substantial momentum, requiring many generations to alter course without causing traumatic disruption.

---

* To be clear, when I refer to the "design" of the body or other ecosystems, I am not implying any designer other than evolutionary processes.

For the past several hundred thousand years, our human ancestors looked like us, were as smart as we are (or maybe smarter, as their brains were about 10 percent larger than ours), and lived in complex, deeply intimate social groups. A number that large can be difficult to comprehend, especially when talking about time. Before I started studying evolution, I was under the impression that ancient Greece was a long time ago. It's "ancient," after all. But ancient Greece was only about three thousand years ago. Rome, two thousand. The very earliest signs of agriculture and fixed human settlements appeared roughly ten thousand years ago. These are all very recent developments, in the context of how long we've been around this place.

A Spanish proverb holds that "habits begin as cobwebs, but end as chains." In evolutionary terms, the mechanics of human self-domestication are complicated. A human baby is a blankish slate upon which novel appetites and beliefs can be etched by society and circumstance: Are women due respect equal to that afforded to men or merely how men have sons? Are dogs lovable pets or a source of meat? Is sex an innocent pleasure or a shameful vice? Whatever our cultures instill in us is written on a slate that's already shaped by biology. It's been grooved, cracked, and colored by hundreds of generations of ancestors who shared common experiences and reactions too numerous to imagine. Cooking food smells good. A loving touch is comforting. Thunder is awesome. Children are precious, and farts are funny.

– Of Capacities and Tendencies –

When talking about human nature, it's crucial to appreciate the difference between *capacities* and *tendencies*. Tendencies can be ignored and overcome, but many capacities are immutable. We can ignore the human tendency to fear the ocean, but we cannot overcome our incapacity to breathe water. We can choose to be vegetarians, but we can't choose to be herbivores. No matter what, we're still omnivores. Whatever choices we make, they are made within the context of our innate, species-specific nature. Human beings are clearly *capable* of a wide range of behaviors, but not all of them resonate equally well with our nature as a species. At least some human beings are *capable* of surviving in extended isolation, for example, but as an intensely social species, this is clearly not in alignment with our nature. Solitary confinement, after all, is a punishment reserved for the worst of the worst.

Although we're demonstrably *capable* of surviving for three score and ten years on a burger-and-beer-based diet and a sedentary lifestyle, we're likely to suffer along the way from tooth decay, obesity, heart disease, diabetes, cancer, and many other ailments. Taking a step into the absurd, we're *capable* of walking backward, but our bodies are clearly made to walk forward. In 1989, an Indian man named Mani Manithan decided to walk only backward when several acts of terror shocked him into thinking his contra-ambulation (initially, naked) would lead to world peace. Mani didn't eliminate global violence, but he proved that it's possible for a human being to spend twenty-five years walking only backward. Still, no one, not even Mani, would argue that it's resonant with our nature.

It's *possible* for human beings to spend their waking lives sitting in cubicles working at thankless jobs under fluorescent lighting, but we shouldn't be surprised by the depression, anxiety, addictive behavior, and sudden explosions of violence such conditions often provoke. Many of us are clearly *capable* of inflicting great pain on others (especially those of us who've suffered abuse at a young age), but nobody suffers from PTSD because they helped a stranger.

Jean Liedloff framed this point in terms of innate "expectations" in her underground classic about hunter-gatherer parenting, *The Continuum Concept*. Describing the evolutionary process of all life-forms, Liedloff wrote, "The design of each individual was a reflection of the experience it expected to encounter. The experience it could tolerate was defined by the circumstances to which its antecedents had adapted." Applying this principle to human beings is "tricky," but it is wise to accommodate these evolved expectations. "[Man's] lungs not only have, but can be said to *be*, an expectation of air, his eyes are an expectation of light rays of the specific range of wavelengths sent out by what is useful to him to see at the hours appropriate for his species to see them. His ears are an expectation of vibrations caused by the events most likely to concern him, including the voices of other people; and his own voice is an expectation of ears functioning similarly in them."

Linguist Daniel Everett spent over twenty years living with the Pirahã, a group of foragers in the upper Amazon. In *Don't Sleep, There Are Snakes*, his memoir of those years, Everett writes, "The Pirahã laugh about everything. They laugh at their own misfortune: when someone's hut blows over in a rainstorm, the occupants laugh more loudly than anyone. They laugh when they catch a lot of fish. They laugh when they catch no fish. They laugh when they're full and they laugh when they're hungry." The laughter of the Pirahã suggests an easy harmony with the world they inhabit—which is the world their minds and bodies expect, because it is essentially the same environment that created them. I don't mean this in a metaphorical sense; I mean it literally. The people Everett describes are at home in the Amazon jungle in the same way a cactus is at home in the desert. Their lives aren't easy, but the difficulties and dangers they face are familiar, because they were faced by many previous generations. You and I, however, live in a world no other human generation has ever known. No wonder so few of us are truly comfortable in the here and now; we've never had a chance to get to know the place.

– A People's History of Prehistory –

Homō hominī lupus est. (Man is wolf to man.)

Any way you look at it, well over 95 percent of the time that our species has existed we've lived as nomadic hunter-gatherers moving about in small bands of 150 or fewer people. Despite significant variation in ecological context, anthropologists have noted near universalities in the behavior and social organization of foragers, from the Amazon basin to the Arctic to the Australian outback. Three characteristics consistently found in foraging societies roughly align with *social, physical,* and *psychological* realms: egalitarianism, mobility, and gratitude. Other aspects of hunter-gatherer life can be seen as extensions of these essential qualities, which anthropologists and ethnographers agree to be ubiquitous among practically all foragers. We'll get into more detail on the specific variations within foraging life later, but for now this will give you a general sense of the foundational principles of our ancestors' social, physical, and psychological lives:

Fierce egalitarianism/sharing. Anthropologists refer to foragers as "fiercely egalitarian," meaning that an individual's autonomy is non-negotiable. Leadership cannot be imposed and tends to be informal and noncoercive, growing out of respect and consensus. Individuals can and do attempt to persuade one another, but they have little or no leverage to impose their wishes. Reciprocity is expected, and the hoarding of food or selfishness of any kind is not tolerated. Children are respected as autonomous individuals and cared for by unrelated adults as well as their biological parents.

Mobility/fusion-fission. Base camps shift and reconfigure frequently, often seasonally, as individuals and small groups move through the environment in search of food. This mobility is an important factor in social organization in that people can easily walk away from uncomfortable situations. Because bands join and split apart frequently, moving to a neighboring band is always an option to avoid brewing conflict or just for a change in social scenery.

Gratitude. Foragers tend to see themselves as the fortunate recipients of a generous environment and benevolent spirit world. The land is the source of all they need. This view is roughly the opposite of the NPP, with its depiction of the natural world as hostile, dangerous, and begrudging. Similarly, foragers tend to relate to a spirit world populated with multiple generous (if sometimes capricious) entities ranging from dead ancestors to elements of their surroundings (water, sky, wind, and so on) rather than the single jealous, vengeful deity at the helm of monotheistic religions.

Until the radical transformations triggered by agriculture around ten thousand years ago, human lives were characterized by *egalitarianism, mobility, obligatory sharing of minimal property, open access* to the necessities of life, and a sense of *gratitude* toward an environment that provided what was needed. In foraging societies, *leaders* were simply those whose opinions happen to be more highly regarded than the views of others at the moment. *Power* was fluid, not something that could be seized, inherited, or purchased. It bears repeating that these features of hunter-gatherer life (representing well over 95 percent of the existence of *Homo sapiens*) are not controversial among scholars, as E. O. Wilson explains in *The Social Conquest of Earth*:

> Hunter-gatherer bands and small agricultural villages are by and large egalitarian. Leadership status is granted individuals on the basis of intelligence and bravery, and through their aging and death it is passed to others, whether close kin or not. Important decisions in egalitarian societies are made during communal feasts, festivals, and religious celebrations. Such is the practice of the few surviving hunter-gatherer bands, scattered in remote areas, mostly in South America, Africa, and Australia, and closest in organization to those prevailing over thousands of years prior to the Neolithic era.

(Note that Wilson specifies "hunter-gatherer bands and small agricultural villages" as being egalitarian. For the sake of simplicity, I'll refer to egalitarianism as being primarily a characteristic of preagricultural societies, but for the sake of clarity and accuracy, it's important to note that the archaeological record makes it clear that egalitarian social organization didn't necessarily dissipate immediately upon the adoption of agriculture. Many small-scale farming communities appeared to perpetuate their egalitarian social organization well after shifting from nomadism. So, while agriculture ultimately led to hierarchical social structures wherever it appeared, the process sometimes took generations.)

The egalitarianism of foragers extends to women as well as men. In *Women in Prehistory*, for example, Margaret Ehrenberg is clear that "social organization is based on equality between individuals and between the sexes. Everyone has equal opportunity to put forward suggestions and have them listened to and every individual has the right to make his or her own decision about what to do in any particular instance."

There are, of course, exceptions to each of these *near* universalities. Like *Homo sapiens sapiens* in every other social configuration, foragers are complicated and variable. A few foraging societies have been documented in which women are treated poorly, child abuse has been reported in several, and egotistical fools manage to wield disproportionate influence and power in others. But such cases are exceptional and often open to questions over whether the people being described are properly classified as "foragers" in the first place. And even the most remote societies have long been affected by encroaching civilization—in the form of contagious disease, air and water contamination, logging, ecological changes affecting hunting and fishing, the disappearance or sudden aggression of neighboring tribes, and so on. But with that caveat, here are some things we can safely say about the origins of our own species:

- Almost all of our ancestors lived as foragers from millions of years ago until the advent of agriculture, roughly ten thousand years ago.
- Foragers living in different ecological circumstances around the world organize their lives in strikingly similar ways. Anthropologists have noted commonalities in how they handle parenting, distribute political power and wealth, resolve conflicts, view relations between the sexes, experience spirituality, and even how they understand and confront death.
- These commonalities exist among disparate societies— isolated from one another—because they arise from foraging itself. These behavioral and cognitive patterns persist because *they work* in the social world of foragers—whatever the physical environment.

- Finally, because these cognitive and behavioral patterns have been integral to the experience of our species for hundreds of thousands of years, they live on within us, having shaped our basic social and political disposition.

Some readers will accuse me of romantic nostalgia and cherry-picking evidence. This is understandable. Reflexive dismissal of any positive view of precivilized life is typical of the civilized, unsurprisingly. Any argument concerning human nature will be a picked-cherry pie. I've included copious references and recommendations for further reading in the endnotes to keep this text from getting too bogged down. No doubt, the myth of the golden age is widespread and is partly fueled by a psychodynamic yearning for the carefree innocence of infancy. But that doesn't delegitimize the need for a critical examination of the present in light of a clearer understanding of the past. Despite the ubiquity of the belief, there is no solid reason to think that things just keep getting better as time goes by—and the psychological motivations underlying blind faith in progress are just as obvious and misleading as any impulse toward nostalgia.

– Noble Savages, Savage Noblemen, and Straw Cave Men –

Subsistence-level hunters aren't necessarily more moral than other people; they just can't get away with selfish behavior because they live in small groups where almost everything is open to scrutiny.

—Sebastian Junger,
Tribe: On Homecoming and Belonging

The characteristic generosity and hospitality of foragers are not fictions concocted by deluded Rousseauian romantics and patchouli-scented hula-hoopers. Unpredictable environments generate challenges best met through reciprocal generosity and hospitality. To the extent that "savages" are noble, it's because they evolved in social groups that cultivate and celebrate generous, respectful behaviors as a means of risk mitigation and self-preservation. If these behaviors have become "innate," they are as innate to you and me as they are to any "savages."

An illuminating tangent reveals that the phrase "noble savage" was born of meaningful confusion and bad intentions. The confusion apparently arose from the two associated meanings of the word "nobility," which connotes both exalted behavior and elevated economic class. Contrary to popular impression, Jean-Jacques Rousseau did not originate the phrase. He never even used it in his writing. In *The Myth of the Noble Savage*, historian Ter Ellingson explains that Marc Lescarbot, a French lawyer-ethnographer, coined the term in 1609, over a century before Rousseau's birth. Lescarbot described American Indians as "truely noble, not having any action but is generous, whether we consider their hunting, or their employment in the wars." Ellingson argues that "the nobility of the Indians is associated with moral qualities such as generosity and proper and fitting behavior." He continues, "The 'savages' of America occupy a status that corresponds, from a legal standpoint, to the nobility of Europe." In other words, Lescarbot saw reflected in the Indians' lives the freedoms, privileges, and responsibilities of the European nobility.

Ellingson explains that the phrase virtually disappeared for about 250 years, when it was resurrected in 1859 by John Crawfurd, a white supremacist who was attempting to become president of the Ethnological Society of London. Crawfurd was disdainful of the emerging anthropological advocacy of universal human rights. He introduced the phrase in a major speech to the society—including the misattribution to Rousseau—as a way to ridicule those who sympathized with such "less advanced" cultures. "Crawfurd's version," writes Ellingson, "becomes the source for every citation of the myth by anthropologists from Lubbock, Tylor, and Boas through the scholars of the late twentieth century." A rhetorical cheap shot from the beginning, the "noble savage" is surely among the longest-lived straw men of all time—still polarizing debate and obstructing nuanced discussion of hunter-gatherer life today.

In *Cannibals and Kings*, anthropologist Marvin Harris explained why Lescarbot may have recognized nobility among the Indians he visited: "In most band and village societies before the evolution of the state, the average human being enjoyed economic and political freedoms which only a privileged minority enjoy today. Men decided for themselves how long they would work on a particular day, what they would work at—or if they would work at all. . . . Neither rent, taxes, nor tribute kept people from doing what they wanted to do." Such relaxed, unconstrained lives would have been striking to a European accustomed to living in a society in which only nobility enjoyed anything similar.

The Indians' wealth—measured in freedom and autonomy—was accompanied by what looked like material poverty to a European. Imagine you are part of a band of fifty or sixty people. Some kids, some old people. Your nomadic band shifts camp regularly. How much stuff do you really want to carry around? If your people use clay pots for cooking, would everyone carry his or her own pot, or would it make more sense to bring a few shared pots? Even the best hunters aren't consistently successful. Most of the time any given hunter will return empty-handed. But when a man kills a deer, what's he going to do with it? Will he share it with only his "wife" and their children, as most evolutionary psychologists claim? Not bloody likely! Such selfishness would lead to social upheaval, potential banishment from the band, rotten meat, and spoiled friendships. The African villager who told my wife that "the best place for extra food is in my friend's stomach" knew his friends would store their extra food in his. When you're living with just enough, as all foragers do by definition, your only insurance policy is the generosity of the people around you. You pay into that policy by being a reliable source of assistance yourself. In this context, it's no surprise that psychologists have established that one of the best ways to improve your sense of well-being is by helping others. It's part of the human design—an important part that has been essential to the survival of our species. Nothing noble or savage about it. Over the millennia in which we became human, a reputation for generosity was important to a successful, happy life. Marvin Harris explains just how traumatic the shift from the egalitarian autonomy of foraging to the coercive power structures of civilization was for our species:

> With the rise of the state, ordinary men seeking to use nature's bounty had to get someone else's permission and had to pay for it with taxes, tribute, or extra labor. . . . For the first time there appeared on earth kings, dictators, high priests, emperors, prime ministers, presidents, governors, mayors, generals, admirals, police chiefs, judges, lawyers, and jailers, along with dungeons, jails, penitentiaries, and concentration camps. Under the tutelage of the state, human beings learned for the first time how to bow, grovel, kneel, and kowtow. In many ways the rise of the state was the descent of the world from freedom to slavery.

But is the egalitarianism of foragers any more aligned with the expectations of human nature than the decidedly unequal access to power, status, and resources people in agricultural societies have faced? It's true that we need to be taught how to be human beings, but some lessons come to us more readily than others. Take the selfish-to-sharing continuum, for instance. On an individual level, we all feel selfish urges. Any organism wants to survive and prosper, which gives rise to immediate me-first impulses. And our species, like the chimps and bonobos to which we're so closely related, is instinctively hierarchical and, being hunters, not unfamiliar with violence.

In many respects, a foraging group is like a soccer or basketball team. If one player is a particularly effective scorer (hunter, gatherer, conflict resolver, storyteller, healer, or decision-maker), this skill is encouraged and admired as long as it results in a net benefit to the group. But once those talents curdle into pride, bossiness, or entitlement, well-established leveling mechanisms get triggered. Such behaviors are discouraged while an egalitarian approach is reinforced through modeling selfless behavior to children and expressing admiration for generosity. If necessary, humor and ridicule can come into play. If making fun of the person doesn't work, social isolation and even death await those who place their own ego gratification above the collective good of the band. When each member of the group depends on the generosity and goodwill of the others—and lethal weapons are always within reach—hearts must remain cool.

In large-scale "civilized" societies, however, we receive conflicting messages on what constitutes proper behavior: Collegial generosity is encouraged on playgrounds and in elementary schools—where you can't eat your candy if you didn't bring enough for everyone—but in business schools and boardrooms, take-no-prisoners competition, acquisition, and individual success tend to be celebrated. Our lives are largely defined by deeply felt conflicts between the reflexive generosity of our hunter-gatherer nature and the inducements to selfishness characteristic of civilization. We all agree that it takes a village to raise a happy child, but most of us ignore or avoid our neighbors and their kids. About 40 percent of Americans donate less than 2 percent of their earnings to charities of any kind, and 45 percent give nothing at all—despite the fact that people who are habitually generous to others are demonstrably happier than miserable misers. We talk, and often think, as if we owned our spouses and children. My wife. My kid. Baby, you belong to me. In a hunter-gatherer band, anyone with such ideas would be seen as a frightening, dangerous, socially inept lunatic facing banishment, or worse.

What's ultimately most striking about the lives of foragers is how utterly familiar they are in ways that still matter. Over thousands of generations, prosocial cultural values have been woven into the biological fabric of our species. These behavior patterns (and the values underlying them) required mutual aid and brains capable of making complex moral judgments—including when and how to punish those who presented a danger to the band.

These fundamental components of human culture evolved along with the biology of our species. Just as our unusually short colon and dull teeth reflect the fact that our ancestors have been cooking their food for a million years or more, our brains reflect, recognize, and reward the social values of a species that survives as a community.

After all, it is in community that our species finds its strength and survival. As individuals, *Homo sapiens* aren't very impressive: weak, slow apes who don't stand much of a chance against an irate raccoon. But bring a few of us together with our communally developed weapons, and we'll bring down a cave bear or woolly mammoth.

This prosocial survival impulse manifests today in our hunger for justice, the quiet comfort we feel sharing food with others, our uncalculating, reflexive feelings of love and protectiveness for children, and the deep relaxation we feel staring into a small fire.

No wonder author Christopher Benfey, in his survey of utopian communities around the world, found that even when separated by time, nationality, and religious orientation, they almost always share a few basic foundational ideas: "that society should be based on cooperation rather than competition; that the nuclear family should be subsumed into the larger community; that property should be held in common; that women should not be subordinate to men; that work of even the most menial kind must be accorded a certain dignity." Not coincidentally, this is essentially a description of the social world in which *Homo sapiens* evolved. Modern humans are lost, and we're looking for ways to go home. The NPP has it backward. Prehistoric man wasn't wolf to man. In fact, he lived in a pretty doggy-dog world.

Chapter 2

Civilization and Its Dissonance

– The Empirical Strikes Back –

There is hope; though not for us.

—Franz Kafka

A few years ago, our family dog got spooked by the Fourth of July fireworks and jumped the fence. Tess was long gone by the time my parents returned from the celebrations. After we had driven around the neighborhood calling her name for hours, the search widened to putting up flyers with her photo (not much help, of course, as all golden retrievers look pretty similar). My parents called various animal shelters and alerted the neighbors. A few desperate days later, a shelter called to say they had a dog matching Tess's description. My sister and father drove down and, sure enough, there she was. After a lot of hugs and face licks, they signed the paperwork and brought her home. Dad, Sis, and the dog were in the backyard, happily tossing a saliva-soaked tennis ball, when my sister's husband came home, looked out the window, and asked my mother, "Whose dog are they playing with?"

He immediately recognized that the dog they were playing with wasn't Tess. He saw what they couldn't because he didn't care as much as they did. You might think the people who knew her best would be most likely to notice that this dog wasn't Tess, but their desperate hope that they'd found their dog blinded them to the fact that they hadn't. It happens to all of us. Our subconscious guides our waking lives just as it does our dreams.

It's emotionally difficult to question progress, because we're so invested in the belief that things are getting better. This tendency serves us well as a survival mechanism, and nobody wants to believe we've invited our children to a party that's already well into the overflowing-ashtrays-and-spilled-drinks phase. But understandable as this optimism may be, we shouldn't mistake it for rational thought.

How often are we encouraged to "never give up hope," when clinging to hope only drags us deeper into hopeless situations? Deluded hopefulness is nourished by a culture that encourages blind faith in progress, no matter how hard this faith bumps up against reality. We Americans are told it's downright unpatriotic ever to awaken from the American dream that anything's possible with enough dedication, focus, and hard work. Every book (including this one) is supposed to end with a hopeful chapter containing five simple steps to everlasting happiness, tighter abs, better orgasms, smarter kids, or financial security. Climate scientists have been warning for decades that we were approaching a "point of no return," but few are willing to announce that we've passed it. Long into the night we reassure ourselves that the sun hasn't quite set. Tobias Wolff wrote that the words "it's never too late . . . still sound to me less like a hope than an epitaph, the last lie we tell before hurling ourselves over the brink."

Of course we all want to believe things are getting better, that our species is learning, growing, and prospering. But what if that's simply not the case? What if all our genuflecting before hope and progress is masking the reality of a situation that is, in fact, already dire and steadily getting worse? In his 132-page gut-punch, *A Short History of Progress*, Ronald Wright explains that "hope drives us to invent new fixes for old messes, which in turn create ever more dangerous messes. Hope," he continued (in 2004), "elects the politician with the biggest empty promise; and as any stockbroker or lottery seller knows, most of us will take a slim hope over prudent and predictable frugality." Wright points to the mythic power of the "religious faith" Western civilization insists we hold toward progress: "Our practical faith in progress has ramified and hardened into an ideology," he wrote. "Progress has an internal logic that can lead beyond reason to catastrophe."

Rosy declarations of eternal progress are as intellectually baseless as they are emotionally comforting, and they undermine our capacity to correct course before it's too late. When you wake up smelling smoke, "Don't worry, go back to sleep" may be precisely what you most want to hear, but that doesn't make it good advice. Psychologist Tali Sharot calls this blind faith in progress "optimism bias." She's found that we tend to dismiss disturbing evidence as aberrations while accentuating anything that paints a brighter picture of the future. Sharot suggests there may be an evolutionary advantage to being hardwired for hope, but Wright sees things differently. At the conclusion of his survey of past civilizations—each of which rose to frighteningly familiar heights of grandeur and avarice before collapsing—Wright warns that we are stumbling past the point of no return. Right now, we have the technological and economic means to alter course, but if we don't seize the moment, "Our fate will twist out of our hands. And this new century will not grow very old before we enter an age of chaos and collapse that will dwarf all the dark ages in our past."

Paeans to progress will always be a part of the civilizational package, because any system predicated upon incessant growth will insist on defining all movement as movement forward, like the falling man who insists he's flying. Until he isn't.

None of this is to say there's been no true progress. There are, unquestionably, aspects of modern life that are significantly better than what came before. But how long before? And what price has been paid for these improvements?

Despair darkens ever more lives as rates of clinical depression and suicide continue their grim climb in the developed world. A third of all American children are obese or seriously overweight, and 54 million of us are prediabetic. Preschoolers represent the fastest-growing market for antidepressants, while the rate of increase of depression among children is over 20 percent annually in recent years. Twenty-four million American adults are thought to suffer from PTSD—mostly attributable to the never-ending wars that have become part of modern life for the swelling underclass with few other employment opportunities. True, we produce more food than ever, but the nutritional quality is suspect at best, and hunger and malnutrition are common in most of the world, while the most fortunate stuff ourselves quite literally to death. Skeletal remains confirm that neither famine nor obesity were common until the advent of civilization.

Modern dentistry? We'll see that the cavities and gum diseases so many of us suffer from didn't arise until the advent of grain-based diets of civilization and monoculture. Scientists analyzing remains from modern-day Sudan found that less than 1 percent of the hunter-gatherers living in the area suffered from tooth decay. Once they adopted agriculture, the rate shot up to around 20 percent.

Most of the dangers civilization claims to protect us from are, in fact, created or amplified by civilization itself. Pointing to antibiotics and bypass surgery in this context amount to extolling the virtues of seat belts and air bags without mentioning that our ancient ancestors were in no danger of auto collisions. If you've set my house on fire, don't expect me to be grateful when you show up later with a bucket of water.

If it's making us unhealthy, unhappy, overworked, humiliated, and frightened, what's all this progress really worth? We know more or less what it costs: nearly everything. We can tabulate the forests destroyed, topsoil eroded, fisheries depleted, aquifers fouled, the atmosphere pumped full of carbon, the cancers, the stress, the desperate refugees, and more. People used to talk about leaving a better world for their children. Now we just hope they'll somehow survive the mess.

The NPP claims that our cleverest ancestors "invented" farming technologies in order to make their lives better. As Jared Diamond explains, "We are accustomed to assuming that the transition from the hunter-gatherer lifestyle to agriculture brought us health, longevity, security, leisure, and great art." But, Diamond notes, "the case for this view *seems* overwhelming, [but] it's hard to prove." In fact, the transition to agriculture was *detrimental* to the overall quality of life for the people born thereafter. Health, longevity, security, and leisure all declined for almost everyone—including, by most relevant measures, the elites.

– Through an Unremembered Gate –

Through the unknown, unremembered gate
When the last of earth left to discover
Is that which was the beginning . . .

—T. S. Eliot, *Four Quartets*

A question I'm often asked: "If agriculture was so bad, why did they choose it?" It's a good question. I wish we could ask Brian Stevenson.

One early-winter morning in 2003, a group of tourists gathered in the parking lot of the Domaine Chandon winery in Napa Valley, California. They'd come for a hot-air balloon flight over the vineyards. As the balloon was being prepared, a sudden breeze kicked up and one of the tourists, a young man from Scotland named Brian Stevenson, grabbed the basket, trying to help out. But the balloon broke free and began to lift off. The professionals knew to let go immediately, but Stevenson hung on as the balloon ascended to several hundred feet over the parking lot, where his grip finally failed, and he fell to his death.

"We have no idea why he held on," the local sheriff said later.

Really? Don't we all know why Brian Stevenson held on? Once his feet were off the ground, he was caught in a loss aversion loop from which the last chance to escape was always already gone. The transition from lending a hand, to holding on for dear life, to the soaring realization that the holding on may have been a fatal mistake probably took no more than a few seconds, but I'd bet that every one of those seconds Stevenson was thinking: "I should've let go before. It's too late now."

Haven't we all been caught in such traps? Who hasn't been in a situation that seemed to make sense at the time, but that ultimately made no sense at all? Who hasn't been mired in a toxic relationship with someone we love too much to leave right now, tonight? Or stuck in a job that scorches the soul but that we can't afford to quit, so we buy expensive toys to mask the pain, thereby making the job even harder to quit?

Once you get a sense that agriculture was not a boon to our ancestors, it's logical to wonder why they chose to abandon foraging in favor of farming in the first place. But that's just it: Our ancestors didn't *choose* to abandon foraging in favor of agriculture any more than Brian Stevenson *chose* to float away from his wife and friends that foggy morning in Napa. In the course of a normal day we take innumerable forgotten steps through unremarkable doorways. Only in retrospect does it sometimes become clear that one of those unremembered gates was in fact a point of no return. One minute you're just hanging out. The next, you're barely hanging on.

The advent of agriculture seems to have been less a clever advance than a desperate attempt to survive. While civilization is generally seen as the result of an unusually stable, benign environment that allowed humanity to benefit from living in complex, highly dense societies, researcher Nick Brooks sees the development of civilization as "an accidental by-product of unplanned adaptation to catastrophic climate change." Civilization was "a last resort"—a response to deteriorating environmental conditions. Our ancestors didn't abandon a desperate foraging existence for the comforts of domesticity. Far from a bold step into a better life, agriculture was a tragic, stumbling misstep into a hole we've been hard at work digging deeper, century by century, as global population exploded far beyond the point of no return.

Jared Diamond's 1999 essay about the transition to agriculture is called, ominously, "The Worst Mistake in the History of the Human Race." More recently, historian Yuval Noah Harari goes so far as to call the agricultural revolution "history's biggest fraud." In his 2015 bestseller, *Sapiens: A Brief History of Humankind*, he writes, "The Agricultural Revolution certainly enlarged the sum total of food at the disposal of humankind, but the extra food did not translate into a better diet or more leisure." Harari agrees that all that extra food merely fueled "population explosions and pampered elites" and that farmers typically worked longer and harder than foragers for an inferior diet. Forced into settled communities as a last resort, agriculturalists faced drastic increases in social inequality, much more violence in the form of organized conflict, and self-appointed elites who used monotheistic religion to lock in their power.

Good ideas tend to spread quickly—even among sparsely distributed forager populations. The archaeological literature is full of examples of the rapid diffusion of new ideas ranging from spear-throwers to pottery design to improved flint-knapping techniques. But judging by the archaeological evidence, nobody was particularly eager to adopt farming. It spread from the Fertile Crescent through Europe more slowly than an old man in slippers, advancing barely a thousand yards per year.

Daniel Everett was struck by the absolute lack of interest among the Pirahã in joining the modern world. On the contrary, they were convinced that he was lucky to be living in theirs. When he asked them if they knew why he was at their village in the Upper Amazon, they replied, "You are here because this is a beautiful place. The water is pretty. There are good things to eat here. The Pirahãs are nice people."

And yet we have all heard the NPP's dark warnings about the brutish existence endured by people like the Pirahã. "It was dangerous." "It was uncomfortable." "Nobody lived past thirty." Civilization has been churning out absurd yet effective warnings against the satisfactions of pretty water, good food, and nice people for millennia. The NPP inflates the value of civilization and demands a knee-jerk rejection of the simple and eternal truths contained in the Pirahã's perspective on life.

In 1929, in *Civilization and Its Discontents*, Freud elucidated the conundrum of the civilized: "Men are beginning to perceive that all this newly won power over space and time, this conquest of the forces of nature, this fulfillment of age-old longings, has not increased the amount of pleasure they can obtain in life, has not made them feel any happier." In the 1920s, when Freud wrote those words, anthropology, sociology, and psychology were all in their infancy, and thus it was very difficult to have any data-based sense of whether our species had lost a sense of well-being or if we'd ever felt it at all—other than as a distant memory of infancy, perhaps. But in the decades since Freud, accumulating evidence has shown that foragers almost never join civilization willingly, and they flee it as rapidly as they can—even when it means retreating into the harshest environments on the planet.

The NPP holds that agriculture began in the Fertile Crescent and spread with the speed of a life-enhancing innovation. In fact, agriculture arose independently in at least eight different parts of the world over about five thousand years, from roughly twelve thousand to seven thousand years ago. In addition to the Fertile Crescent, archaeologists have identified evidence of a transition from foraging to farming at sites in the north and south of China, the Andes, central Mexico, New Guinea, Egypt, the Mississippi Valley, and West Africa. There is no evidence that agriculture spread to any of these places from the Fertile Crescent. Rather, it appears that similar sequences of climactic changes triggered the shift to farming.

While agriculture arose independently and variously, its effects on people's lives were overwhelming and universal. Agriculture was far more than just a way of getting food. It shaped practically every element of human societies (male-female relations, child care, government, class system, militarism, humans' relations to other animals and the natural world, and so on). The story changed, and with it, the world.

This is a crucial, often missed, point about the transition from foraging to farming. The change wasn't merely a pivotal point in how our species lived in the world. It marked a fundamental shift in what kind of world human beings inhabited, both materially and conceptually. It isn't hyperbole to say that agriculture extracted humans *from* the world and pitted us *against* it. Niles Eldredge of the American Museum of Natural History has written that the shift to agriculture and resulting civilization removed our species from the relation with the natural world that we had until then shared with every other species since life began. "We abruptly stepped out of the local ecosystem. . . . Our interests no longer dovetail[ed] with those of the natural world around us. . . ." Adopting agriculture was "tantamount to declaring war on local ecosystems."

But how did our species go from being a participant in flow with the natural world to where we are now, hanging on for dear life as we float ever further from an integrated, sustainable relationship? It seems that things got very good before getting very bad—a common, if disastrous, sequence of events, as any gambler or junkie will attest. Until about fifteen thousand years ago, the planet was climactically unstable: polar ice caps spreading and contracting, sea levels rising and falling, extreme climactic fluctuations and sudden shifts in ocean currents. At that point, a massive sheet of ice covered much of the northern hemisphere. All of Scandinavia, northern Germany, and much of Britain was frozen solid. Sea levels were about ninety meters lower than they are today, due to all that water being stored in ice.

Ice core samples taken in Greenland reveal a dramatic shift from these cold, dry, unstable conditions to a new period of markedly warmer temperatures and increased rainfall that lasted a few thousand years—plenty of time for people to get used to the surplus food and for population levels to reach the higher carrying capacity of this wetter, warmer world. Ice sheets retreated, temperatures went up, rains came down, plants bloomed, and animals reproduced. A long summer had begun.

In 2001, Peter Richerson, Robert Boyd, and Robert Bettinger published a powerfully argued paper called "Was Agriculture Impossible During the Pleistocene but Mandatory During the Holocene? A Climate Change Hypothesis." Conditions were so good, they argue, that some foraging societies in the Fertile

Crescent area began to shift toward something like agriculture before anyone actually started farming. Evidence of small settlements dating as far back as fifteen thousand years ago has been found in the area that today stretches from southwest Turkey down through Syria, Lebanon, Jordan, Israel, and the Palestinian territories. Known as Natufian villages, these settlements appear to have held as many as a few hundred foragers each. Archaeologists have uncovered signs of settled villages around rich and reliable food sources, and evidence of more elaborate spiritual practices, which suggest more hierarchical social organization and probably more organized conflict between groups. Thus, some of the ills generally associated with agriculture may have arisen before farming itself in some areas.

We find something similar to these social patterns in so-called complex hunter-gatherers, such as the tribes native to the Pacific Northwest region of North America. Seasonal salmon runs, seal hunts, and whale hunting provided something like a harvest, and by smoking the surplus meat, the Tlingit, Haida, Coast Salish, Chinook, and others could save stockpiles of food for later. Whatever its provenance, accumulated wealth almost always generates political hierarchies, increasingly complex rituals and artistic creation, raiding and warfare, and enslavement.

A few miles north of the town of Urfa, in the Anatolia region of Turkey, lie the extraordinary ruins of Göbekli Tepe, probably built about twelve thousand years ago. Göbekli Tepe was already over six thousand years old when the Great Pyramid was being built, and it is the most ancient megalithic complex ever discovered— by a long shot. Until the discovery and carbon dating of Göbekli Tepe, the oldest megalithic site known was on the island of Malta, estimated to have been built around fifty-five hundred years ago.

The ruins in Turkey contain more than sixty T-shaped limestone pillars, each weighing several tons. Most of the pillars are engraved with bas-reliefs of dangerous animals like scorpions, snakes, boars, and lions. But perhaps the most striking feature of Göbekli Tepe is what it *doesn't* contain. There are no signs of human habitation: no houses, no firepits, no remains of domesticated animals or plants. Since nobody lived there, it stands to reason the temple was built by foragers *before* agriculture took hold in the area. This line of thinking upends established ideas about foragers and the origins of formal religion in that it sees organized religion preceding (and eventually necessitating) agriculture.

Klaus Schmidt, the German archaeologist who discovered the site and led its excavation from 1989 until his death in 2014, promoted this view: "Göbekli Tepe is not a house or a domestic building," he said. "Evidence of any domestic use is entirely lacking. No remains of settled human habitation have been found nearby. That leaves one purpose: religion. Göbekli Tepe is the oldest temple in the world. And it isn't just a temple; I think it is probably a funerary complex." Schmidt believed ancient hunters brought their dead to Göbekli Tepe, where they were picked clean by vultures and other animals, much like the "sky burials" still practiced in Tibet.

Schmidt, himself a Catholic, was convinced that it was the urge to worship together that brought people into the first stable settlements. The construction and maintenance of ambitious temples like Göbekli Tepe, he believed, necessitated the development of agriculture as a way to feed the teams of workers and keep the work moving forward. The gods came first, in Schmidt's view, and they demanded the rest. Possibly. But the climate-based argument articulated by Richerson, Boyd, and Bettinger seems more convincing, making Göbekli Tepe less a trigger of agriculture than an indication that the cultural ground was fertile for such a transition.

In any case, the people who built Göbekli Tepe certainly had plenty of reason to feel grateful. They were living in a nearly ideal environment. The dry and barren hills that now stretch off in every direction looked very different twelve thousand years ago. Food was everywhere. Grasslands with two kinds of wild rye and einkorn wheat carpeted the hills. Interspersed throughout the meadows and grasslands were forests of oak, pistachio, and other nut-bearing trees. Gazelle ranged the area, and local people harvested them en masse—sometimes taking entire herds at once. Aurochs (the wild ancestor to domesticated cattle) were plentiful and often weighed up to two thousand pounds each. Schmidt has described the area as having been "a paradisiacal place." It must have been, to have supplied enough food to maintain the people who worked to build such a temple. "They were having big parties," Schmidt told journalist Elif Batuman, possibly including beer and other, stronger consciousness-altering substances.

The transition into these villages must have been relatively painless. During this long summer that lasted several centuries, life must have been easy and rich. Game was plentiful, the land overflowing with fruits, nuts, and seeding plants. Like the Pirahã, these first villagers must have felt grateful for a world that, while challenging at times, was bountiful and nurturing. The environment was so generous that people no longer had to keep moving to find the next day's meal. They'd settled in the richest valleys and riversides, wandering into the surrounding hills to hunt and gather, or throwing nets into the water to pull out dinner. Complex cultures grew out of this abundance, some of which gathered periodically to trade, to intermarry, to tell stories, and, in the case of Göbekli Tepe, to honor their dead in rituals involving sacred temples that our species had—for perhaps the first time ever—made rather than found.

Since at least thirty-five thousand years ago, people had been painting images of bison, horses, and their own handprints on rock walls, but the people who built Göbekli Tepe were not merely modifying a cave wall by adding some ochre or charcoal; they were constructing their own stone walls by cutting and arranging massive human-shaped blocks that weighed as much as a hundred of the men who struggled to shift them into place.

But all this abundance created a structural vulnerability. In *The Long Summer*, archaeologist Brian Fagan explains that generations of people grew accustomed to living in static villages that could only exist in an extraordinarily rich ecosystem. The flexibility and interdependence of foraging societies fell away as more fixed social systems took hold. "No longer could people simply move away to better-watered locations or fall back on less favored ones" as humans had forever done. People had lost their capacity for mobility, "a social flexibility that was as old as humanity itself."

When disaster struck, it came from the other side of the world. In North America, a massive lake had formed from the meltwaters of the retreating ice sheets. Now known to researchers as Lake Agassiz, this huge body of ice water extended from modern-day Manitoba to Minnesota, covering an area of around 440,000 square kilometers—larger than all the modern Great Lakes put together. Somewhere between 13,500 and 12,600 years ago, Lake Agassiz emptied into the Labrador Sea, rocking the entire planet. The Atlantic Meridional Overturning Circulation—which brought water from the tropics up into the North Atlantic, thus warming Europe—was blocked by the sudden influx of ice-cold fresh water. (A similar process appears to be under way now, as the Arctic ice sheets melt into northern oceans.) Glaciers that had been retreating to the north resumed their icy advance to the south. Severe winter storms lashed Europe with freezing winds that hadn't been felt for several thousand years. The snows of what scientists call the Younger Dryas period dominated the north. Even much farther south, around Göbekli Tepe, temperatures dropped about twelve degrees Fahrenheit. A thousand-year drought began, and the long summer had come to an abrupt end.

Agriculture, then, appears to have been a panic-stricken response to sudden, desperate changes. Population density was too high to adjust to the reduction in available food without massive die-offs. People had grown accustomed to living in villages, and as plentiful wild food sources dried up, more hungry people must have streamed in from the hinterlands. Incipient hierarchies already in place gained new importance and power.

Once Lake Agassiz broke over its banks and flooded into the North Atlantic, the generous gods of plenitude appear to have been ousted by an angry, jealous, vengeful, and stingy god. Where the previous gods had resembled loving parents, this one was closer to a cruel, exploitative slave owner. Under his reign, those who don't work, don't eat. And even if you work from dawn till dusk, you may still go to sleep hungry. The gods of ease and play, pleasure and laughter were out; grueling, backbreaking work was in. We still worship this god of toil, sacrifice, scarcity, and submission. No pain, no grain.

Whether the initial steps toward agriculture involved planting wild seeds closer to water or digging an irrigation channel to bring water to withering nut trees, we know that these steps took our species through an unremembered gate toward modernity. Clever, desperate people were just trying to tweak things a little to produce more food in a time of desperate need. Like Brian Stevenson lending a hand on that foggy Napa morning, they had the best of intentions. But the day our kind first managed to *produce* food rather than find it, their feet left the ground, and it was already too late to let go.

Once begun, the agricultural revolution was a one-way, ratcheting process. But what choice did they have? Only in hindsight has it become clear that in struggling for their own short-term survival, they were taking the first steps down a path that human beings had never trod before, a path that would lead us away from everything we'd been since the origin of our species.

Because farming is so successful in temporarily producing more food per unit of land—often up to a hundred times more than foraging—already overpopulated areas soon swarmed with ever more hungry people. Since farming is labor-intensive work, pools of cheap labor were needed by those who owned the land. The notion of ownership—something that had been limited to a favorite spear, necklace, or piece of clothing—now took on almost magical power. Men could now own not only land, but surplus food and seeds, sources of water, animals, and soon enough, other human beings. Because babies could now be weaned much earlier with milk from domesticated animals, women became pregnant again just a year or two after giving birth—resulting in fertility much higher than that of foragers, who typically breast-fed their children for three or four years before becoming pregnant again.

Women—who had been respected members of egalitarian foraging societies—were now reduced to a status close to that of domesticated animals. As their role shifted from food gatherers to child producers, they had little say in whom they married, how many children they'd have, or any other matter of consequence to their lives. When we read in Exodus 20:17 "Thou shalt not covet thy neighbor's wife," most of us take it as an admonition to respect thy neighbor's marriage. But read in context, it takes a very different tone: "Thou shalt not covet thy neighbor's house, thou shalt not covet thy neighbor's wife, nor his manservant, nor his maidservant, nor his ox, nor his ass, nor any thing that is thy neighbor's." Far from an admonition to respect thy neighbor's marriage, this is all about respecting thy neighbor's property— wife included.

Aside from the exceptional complex hunter-gatherers I mentioned earlier, whose unusual ecological situation mirrors agricultural social structures in crucial ways, no one had ever lived this way before. These densely populated settlements required new social institutions to manage novel complexities. Ownership of land, animals, slaves, and women needed to be codified. The concept of property permeates the Old Testament—as it still does the modern world—which makes it easy to forget how radically new to our species the concept was. Ancient habits of egalitarianism and generosity were abandoned. Nomadism was displaced by a sedentary life allowing for the accumulation of possessions, from goats to wives to children to slaves. This shift represented the abandonment of a way of living that had served our species well since the beginning of time. Everything changed.

In many ways, *Homo sapiens sapiens* became a different sort of animal at this historical watershed moment. From then on, right through today and tomorrow, almost every "civilized" member of our species lived in a social world governed by institutions that demanded behavior often in direct conflict with innate capacities and predilections that had evolved over millions of years—years in which sharing and individual autonomy were essential elements of human survival. Our species went from living in the world to living in a zoo of our own making. *Without understanding what was happening, our ancestors were being domesticated as surely as were their plants and animals.* Along with their domesticated animals, human beings now lived in overcrowded, disease-ridden enclosures full of their own excrement, herded about without explanation or redress, beaten and whipped into compliance, bought, sold, and slaughtered. Wright reminds us that "we call agriculture and civilization 'inventions' or 'experiments' because that is how they look in hindsight. But they began accidentally, a series of seductive steps down a path leading, for most people, to lives of monotony and toil." Yes, farming brought more food, but it was far less nutritious food. Human population exploded while quality of life collapsed. Civilization is like a hole our clever species dug and then promptly fell into.

The influx of people, along with annual cycles of surplus and scarcity (harvest and waiting for the next harvest) *required* the rapid development of strictly enforced hierarchies and specialization of labor. Priests and rulers were needed to direct and exploit the labor of commoners. Guards had to be retained and paid to protect the year's harvest, enforce planting schedules, and run down thieves. Soldiers had to protect the accumulated wealth of settlements from raids—or to conduct raids to steal the wealth of other settlements. Economic disparities widened between those who owned the land and animals and those who had nothing to sell but their own time, sweat, and suffering.

Many types of conflict now became inevitable, both within these settlements and between them. The much higher fertility rates of farming people meant that ever more land was needed to feed growing populations. Thus was born the rapacious god of economic growth who continues to rule today. Growing economies spread aggressively, first to provide land for new generations of farmers and then to replace the formerly fertile lands swept away by rains when the forests had been cut down for fuel and could no longer slow the loss of soil. Inherently expansionist agricultural societies consumed and exhausted the land, then spread out to conquer and occupy more. "Savages" and "barbarians" were efficiently exterminated or driven off, and the cycle began anew.

We'll never know for certain what inspired those last generations of foragers to build Göbekli Tepe, but their descendants appear to have had serious regrets. These extraordinary temples were a celebration of an extraordinarily abundant (dare we say

Edenic) period that, counterintuitively, triggered the long poverty of civilization. They mark the transition from our species' long nomadic past of egalitarianism, autonomy, and gratitude into a world of owners and the owned. Perhaps this explains why the temples weren't merely abandoned by the descendants of the people who'd built them. Despite their hunger and desperation, they went to the trouble of burying Göbekli Tepe, in garbage. The first sacred temple ever built by human beings ended as a dump.

– "The Best People in the World" –

There are many humorous things in the world; among them, the white man's notion that he is less savage than the other savages.

—Mark Twain, *Following the Equator*

A central theme of the NPP is the notion that *we* are more advanced, cultured, sophisticated, chosen, and evolved than them. We are civilized. Our superiority is self-evident, and evidence to the contrary tends to fall through the cracks of history.

Upon his first encounters with the native people he "discovered" in the West Indies, Columbus was struck by their kindness, generosity, and physical beauty. In a letter to the king and queen of Spain, he explained: "They are very simple and honest and exceedingly liberal with all they have, none of them refusing anything he may possess when he is asked for it. They exhibit great love toward all others in preference to themselves." In his own journals, he was even more complimentary: "They are the best people in the world and above all the gentlest—without knowledge of what is evil—nor do they murder or steal . . . they love their neighbors as themselves and they have the sweetest talk in the world . . . always laughing." A few pages on, in one of the most chilling pivots in recorded history, Columbus wrote: "They would make fine servants. With fifty men we could subjugate them all and make them do whatever we want."

The word "gold" appears seventy-five times in just the first two weeks of Columbus's journal entries. The celebrated explorer's obsession with the accumulation of gold led to a hellish system in which Indians who failed to deliver their assigned quota of gold had their limbs hacked off. That there was little gold to be found on these islands didn't matter to the Europeans. As the otherwise admiring Columbus biographer Samuel Eliot Morison admits, there was no escape from the maniacal Europeans: "Those who fled to the mountains were hunted with hounds, and of those who escaped, starvation and disease took toll, while thousands of the poor creatures in desperation took cassava poison to end their miseries." Morison estimates that a third of the three hundred thousand Taíno perished in just two years, from 1494 to 1496, and by 1508, only sixty thousand survived. Within a few decades, only a few hundred of "the best people in the world" were left.

I visited Casa de Colón, a museum devoted to Christopher Columbus, on the island of Gran Canaria. Lodged in a building Columbus supposedly stayed in when he stopped over on his voyages, the only mention I could find of interactions with native people was a re-creation of a Vatican document specifying that the Taíno were to be treated well—surely one of the most profoundly ignored proclamations ever. It occurred to me that this omission was like a Hitler museum failing to mention the Holocaust.

What happened to the Taíno was just a taste of the genocide that awaited the native people of the New World. By 1600, over 90 percent of the native population of the Americas was gone, a staggering fact Ronald Wright has called "the greatest mortality in history." Around 56 million people died in South, Central, and North America in the hundred years following first contact with Europeans. So many were lost, in fact, that the ecological changes caused by their sudden absence may well have triggered the so-called Little Ice Age experienced in Europe in the early 1600s.

A Dominican priest named Bartolomé de Las Casas witnessed and recorded some of the crimes of the Spanish in *A Short Account of the Destruction of the Indies* (published in 1552). "Of all the infinite universe of humanity, these people are the most guileless, the most devoid of wickedness and duplicity," wrote the priest. "Yet into this sheepfold . . . there came some Spaniards who immediately behaved like ravening beasts." Las Casas wrote of soldiers testing their blades by casually slashing passing Indians, smashing babies' heads against rocks for no reason at all. Any Indians who resisted were hunted down and murdered. Those who stole food were beheaded or burned alive. To this day, native people in the Upper Amazon refer to outsiders as *pishtaco*, which translates roughly to "those who steal your oil." But they're not talking about petroleum. The term is thought to date back to the sixteenth century, when Spanish conquistadors such as Lope de Aguirre first appeared in the area. Some of the Spaniards, seeking a way to keep their iron weapons from rusting in the jungle humidity, were said to have killed native people and boiled down their bodies for fat with which to grease their guns.

In 1550, the Vatican arranged a debate between Las Casas, who represented the rights of Indians, and Juan Ginés de Sepúlveda, who argued that the Indians were not human and thus had no soul or claim to human dignity. Las Casas claimed to have won the so-called debate in Valladolid, Spain, but if so, like the Vatican document on display in Casa de Colón, it was a paper victory. Las Casas wasn't alone in his outrage and shame at the behavior of these Christians. A group of Dominican friars recounted "unspeakable atrocities." They reported that children were being thrown to dogs to be eaten, women raped, men murdered for a laugh. You may be wondering what the hell was wrong with the Spaniards, but their behavior—demonic as it was—was far from unusual for "civilized" explorers of their day. These men hadn't lost their way. This *was* their way.

"The enormities perpetrated in the South Seas upon some inoffensive islanders well-nigh pass belief," wrote whaler-turned-author Herman Melville in a letter to his brother. "These things are seldom proclaimed at home, they happen at the very ends of the earth; they are done in a corner, and there are none to reveal them." In 1910, Anglo-Irish diplomat Roger Casement spent several months among rubber traders in the Amazon. His account of the treatment of native people echoes Las Casas: "These [people] are not only murdered, flogged, chained up like wild beasts, hunted far and wide and their dwellings burnt, their wives raped, their children dragged away to slavery and outrage, but are shamelessly swindled into the bargain."

History is full of accounts like these where the civilized meet the other. Other civilizations have been no less brutal, as the human sacrifices favored by the Aztecs and Mayans, slavery in ancient Rome and several African empires, and the rape and pillage preferred by the Mongol hordes illustrate. Historically, those who see themselves as "civilized" see the noncivilized as less than human and therefore disposable. For the mighty, might makes right.

Just as the Taíno shared core elements of their social, spiritual, and economic experience with other hunter-gatherer people around the globe, the Spaniards who perpetrated their genocide recognized similarities to their own world when they encountered the Aztecs of Mexico and the Incas in Peru. Like the Europeans, both the Aztecs and Incas were hierarchical agricultural empires ruled by delusional egomaniacs who commanded large, highly organized armies with which they dominated and decimated the smaller-scale societies within reach.

When Hernán Cortés arrived in the Aztec capital, Tenochtitlan, on November 8, 1519, he walked into one of the largest cities in the world, with a population estimated at between 200,000 and 300,000 people. In Europe, only Paris, Venice, and Constantinople were comparable in size. The clash between the Spanish and the Aztecs was a clash between civilizations. Columbus's ravaging of the Taíno, on the other hand, was a civilization encountering a foraging society.

It is a mistake to dismiss this brutality as human nature, unless we accept the premise that the Spaniards were more *human* than the Taíno. Were the Aztecs more *human* than their victims? The Romans certainly believed themselves to be qualitatively superior to *barbarians*, but haven't we moved beyond those sorts of assumptions? One of Darwin's greatest (and most controversial) gifts was to provide the scientific evidence that all human beings are equally evolved, in that we all come from common ancestors.

Once we accept that all human beings are, in fact, equally human, it becomes clear that *human nature* offers little to help explain systematic cruelties common to civilizations but rare or nonexistent among foragers (subjugation of women, slavery, extreme disparities in wealth, and so on). What fueled the Spaniards' cruelties wasn't human nature. It was civilization. Civilization convinced the Spaniards that their superior weapons made them superior beings. Civilization created the filthy cities in which their ancestors acquired immunity to the pathogens that wiped out millions in the Americas. Civilization convinced Columbus and his men that gold was more valuable than the lives of the people they destroyed to get it. Civilization twisted their souls into somehow concluding that their savior, supposedly

57

the embodiment of love and mercy, would have approved of—demanded, even—the enslavement, murder, and mutilation of the best people in the world.

None of this is meant to be an indictment of sixteenth-century Europeans and Aztecs, Christians, Silicon Valley entrepreneurs, or any other peddlers of progress—but of a particular form of social organization that we self-congratulatingly call "civilization." Whether in China, Africa, the Americas, India, or Europe, the lives of people in civilizations are strikingly similar in many important respects—as are the lives of hunter-gatherers. "Civilized" people—whether Aztecs or Australians—have always believed that they are better than so-called savages. In fact, while there are thousands of recorded cases of people from civilized communities fleeing to "go native," there are few documented cases of native people willingly choosing to join civilization when they had any other viable options. A truly superior social system would have no need to forcibly recruit new members, but the history of civilization is, as we'll see, replete with systems that enforce participation.

– The Art of Not Being Civilized –

I am convinced that those societies (as the Indians) which live without government enjoy in their general mass an infinitely greater degree of happiness than those who live under the European governments.

—Thomas Jefferson (in a personal letter, 1787)

The Art of Not Being Governed, by James C. Scott, is about people who have tried to slip the yoke of civilization, and the ways civilization pulled them in anyway. Scott, who teaches political science and anthropology at Yale, notes that "much, if not most, of the population of early states was unfree," and that attempts to escape were common. In addition to the disease, famine, and abuse suffered by the civilized, early states were "warmaking machines" that triggered "hemorrhages of subjects fleeing conscription, invasion, and plunder."

In contrast to the servitude of these early states stand innumerable accounts of the autonomy, personal freedom, and satisfaction of foragers. Consider Everett's description of the happiness of the Pirahã: "Since my first night among them," he writes, "I have been impressed with their patience, their happiness, and their kindness. This pervasive happiness is hard to explain, though I believe that the Pirahãs are so confident and secure in their ability to handle anything that their environment throws at them that they can enjoy whatever comes their way."

How comfortable are you that you can handle anything that comes your way? Tax audit? Prison sentence? Unemployment? Me neither.

This autonomy and confidence in the face of what the natural world presents wasn't limited to some idealized "Man the Hunter." Margaret Ehrenberg is clear that "the status of women is regularly higher in forager groups than in any other type," and that "social organisation is based on equality between individuals and between the sexes." Structurally speaking, this gender equality isn't surprising. When every adult has direct access to the necessities of life (food, shelter, community), children are cared for communally, and possessions are few and easily replaced, there are few opportunities for coercive power over others. When cultural values celebrate individual autonomy, respectful sharing of resources, and mutually beneficial interdependence, the logical result is a society in which people are generally satisfied with their lives and not overly concerned with telling others how to live theirs.

The vast majority of anthropologists who've observed foragers reported relative equality between men and women, and the logic of free access to resources and shared responsibility for child care seems to predict it as well. A recent study conducted by Mark Dyble and Andrea Migliano found that gender equality may have played a crucial role in the survival of our species. Building off a 2011 study of thirty-two hunter-gatherer societies that found high levels of unrelated individuals living in the same bands, Dyble and Migliano hypothesized that where men held disproportionate power, families would tend to live with the men's relations, but if men and women had equal say in living arrangements, they'd often cluster with unrelated people. After collecting data for two years among the Palanan Agta people of the Philippines and the Mbendjele of Central Africa, they found that four times as many of the egalitarian foragers were unrelated to others in their living group as compared to a male-dominant farming society living nearby. Living with unrelated individuals fueled the human tendency to cooperate beyond the gene pool, according to Migliano: "In forming mainly unrelated camps, hunter-gatherers evolved the capacity to cooperate with unrelated individuals."

But if necessity is the mother of invention, satisfaction may dampen the appetite for what we call "progress." The subversively unproductive contentment of noncivilized people vexed Darwin. "Nomadic habits, whether over wide plains, or through the dense forests of the tropics, or along the shores of the sea, have in every case been highly detrimental," he wrote in *The Descent of Man*. "The possession of some property, a fixed abode, and the union of many families under a chief, [are] the indispensable requisites for civilisation." Earlier, in *Voyage of the Beagle*, he lamented the "perfect equality" he perceived among the Fuegian tribes, which, he was certain, "must for a long time retard their civilisation."

The apparent abundance of the foraging life offers a stark contrast to the extreme toil and poverty of early farmers, famously described in the book of Genesis: "And to Adam he said, '... cursed is the ground because of you; in pain you shall eat of it all the days of your life; thorns and thistles it shall bring forth for you; and you shall eat the plants of the field. By the sweat of your face you shall eat bread, till you return to the ground, for out of it you were taken; for you are dust, and to dust you shall return." Moving from the comparatively free and easy state of hunter-gatherers to the servitude of a farming life was always difficult and often compelled. It was perhaps the most traumatic transition in the history of our species. The fall from grace.

Historically, in settlements where the surrounding environment offered opportunity for subsistence living, people had to be coerced into joining civilization. Scott describes the brutal subjugation as "anything but a benign, voluntary journey toward civilization." In fact, large portions of these early civilizations were not participants; they were property, "taken en masse as prizes of war and driven back to the core or purchased, retail, as it were, from slaving expeditions selling the state what it most needed." What these early states "most needed" was cheap human labor to keep the wheels of civilization turning: workers to plant and harvest crops, armies to conquer and hold new land, slaves to dig canals and cut roads.

This insatiable hunger for human labor also helps explain why most major religions so insistently and violently oppose nonreproductive sexual behavior—a major source of human suffering in civilized societies. Despite these prohibitions, nonreproductive sex can practically be considered a defining human characteristic. We are one of a very few species that enthusiastically engage in sex in myriad ways that can't possibly lead to pregnancy, but many religions impose draconian punishments for masturbation, sodomy, same-sex dalliances, or even enjoying sex with one's marital partner a little too much or too often. Seen as a way of compelling rapid population growth in order to fuel the growth of civilized populations, this otherwise bizarre prohibition of nonreproductive sex begins to make sense. Humans are in effect being bred as a source of cheap, disposable labor, like horses, oxen, or camels.

Forcing the reluctant to join expanding empires wasn't restricted to biblical or classical times. In *The Invention of Capitalism*, economic historian Michael Perelman explains how the economic noose was tightened around the necks of anyone who tried to opt out of the civilizational enterprise in the early days of the Industrial Revolution. "Rather than contending that market forces should determine the fate of these small-scale producers, classical political economy called for state interventions of one sort or another to hobble these people's ability to produce for their own needs." It wasn't enough merely to *be* civilized yourself; everyone else had to be civilized, too. Perelman quotes a botanist, Thomas Pennant, who explored the Scottish Highlands in the 1760s. His descriptions of the Highlanders are reminiscent of what we've heard many times about native people living beyond the fence: "The men are thin, but strong; idle and lazy, except when employed in the chace, or anything that looks for amusement; and are content with their hard fare, and will not exert themselves farther than what they deem necessaries." Pennant's description resembles Adam Smith's opinions concerning the uncivilized. "The life of a savage, when we take a distant view of it, seems to be a life of either profound indolence, or of great and astonishing adventures."

This state of affairs could not be permitted. Men had to be made poor enough that they'd be forced to join the desperate throngs in the mines, armies, and factories. A London police magistrate named Patrick Colquhoun articulated the widespread view that poverty was integral to the health of civilization: "Poverty . . . is a most necessary and indispensable ingredient in society, without which nations and communities could not exist in a state of civilization. It is the lot of man. It is the source of wealth, since without poverty, there could be no labour; there could be no riches, no refinement, no comfort, and no benefit to those who may be possessed of wealth."

The systematic coercion of those who tried to opt out "cut through traditional lifeways like scissors," explains Perelman. "The first blade served to undermine the ability of people to provide for themselves. The other blade was a system of stern measures required to keep people from finding alternative survival strategies outside the system of wage labor." One of the so-called Tudor Poor Laws, enacted in the late 1500s, outlawed begging in England. Anyone over the age of fourteen caught begging would be flogged and branded with a red-hot iron on the left ear. Anyone caught a third time was to be executed.

These examples are not exceptional. Francis Hutcheson, one of Adam Smith's most important mentors, was one of the leading moral philosophers of his day (mid-1700s). Hutcheson counseled: "If a people have not acquired an habit of industry, the cheapness of all the necessaries of life encourages sloth. The best remedy is to raise the demand for all necessaries. . . . Sloth should be punished by temporary servitude at least."

And make no mistake, people are still being dragged into the market economy. Multinational corporations routinely expropriate land in poor countries (or "buy" it from corrupt politicians), force the local populations off the land (so they cannot grow or hunt their own food), and offer the "luckiest" among them jobs cutting down the forest, mining minerals, or harvesting fruit in exchange for slave wages often paid in company currency that can only be used to buy unhealthful, industrially produced food at inflated prices at a company-owned store. These victims of market incursion are then often celebrated as having been saved from "abject poverty." With their gardens, animals, fishing, and hunting, they had been living on less than a dollar per day. Now, as slave laborers, they're participating in the economy. This, we're told, is progress.

In 2014, the Mehdiganj Coca-Cola bottling plant near Varanasi, India, was shut down by the government after years of protest by local residents. People all over India had been denouncing the company's policy of extracting so much water from aquifers that local wells ran dry. On the other side of the world, in 1999, a division of Bechtel, the secretive American defense contractor with top-level ties to the Reagan and Bush administrations, bought the municipal water system of Cochabamba, Bolivia, from the federal government. Soon, representatives of the company arrived to install meters on wells—many of which had originally been dug and maintained by village cooperatives. The local people saw their water bills surge 50 percent, on average—often for water from wells *they themselves had dug*. They were also expected to pay for the installation of the new water meters and warned that collecting rainwater was now illegal.

From foragers being forced off land they've lived on for centuries because they cannot produce deeds of ownership, to eighteenth-century Scottish Highlanders who preferred to tend their sheep, to today's college graduates saddled with tens of thousands of dollars in debt before they've landed their first job, nonparticipation in the market economy has consistently and effectively been eliminated as a viable option. To those who suggest we should "Love it or leave it," I'd suggest that neither option is—or has ever been—a realistic possibility. It's as if people are being forced into casinos at gunpoint, where they lose everything, generation after generation, and then they're told they've got a gambling problem.

Before turning to how misguided assumptions about foragers and the civilized contaminate contemporary perspectives on the natural world and human nature, let's take a closer look at Malthus and Hobbes, arguably two of the most important thinkers of the past five hundred years.

– Malthusian Miscalculations and Hobbesian Horror Shows –

Thomas Malthus, the world's first professor of economics, is to human misery what Mozart is to powdered wigs. It's perhaps fitting that history remembers him for an idea that is dismal and influential, but utterly mistaken.

By the end of the eighteenth century, the British economy was generating enough wealth that radical thinkers like William Godwin (a close friend of Malthus's father) could suggest that everyone could live comfortably if a more egalitarian approach to wealth distribution were undertaken, an idea enjoying a resurgence today in the form of Guaranteed Basic Income.*

Largely as a challenge to what he considered to be the naïvely utopian views of his father's friend, the thirty-two-year-old Malthus published *An Essay on the Principle of Population* in 1798, in which he argued that there is no point in helping the poor, because if wealth were more equally distributed and everyone had enough to eat, population would increase geometrically, as two parents give birth to four children who would have eight,

---

* Godwin, by the way, deserves to be far better known, as his thinking underlies everything from feminism to *Frankenstein*.

and so on—while food supply could only grow arithmetically, as new land is placed under cultivation. Since population would therefore always grow far more quickly than resources, Malthus argued, scarcity and starvation are simply unavoidable aspects of life. There could never be, and never had been, enough for everyone. From these seemingly irrefutable calculations came the brutal Malthusian dogma that chronic overpopulation and, therefore, crushing poverty will always be and have always been the inescapable fate of human beings. "The poverty and misery which prevail among the lower classes of society," Malthus wrote, "are absolutely irremediable."

While this may strike us as horrible news, it was welcomed by many of those in the upper classes, as it absolved them of any responsibility for the brutal, widespread poverty of their age and gave them a powerful justification for doing nothing to address it. If the situation is "absolutely irremediable," one might as well focus on one's tea and crumpets and not fret over the suffering of the destitute, as the poor will always be with us. And if you can convince people that their dire conditions are the natural and eternal state of affairs, you castrate the impulse for rebellion. Agitating for changes to the natural human condition, after all, would be as pointless as marching against nighttime.

Malthus had based his estimates of human reproductive rates on the growth of European populations in North America over the previous 150 years. Noting that the colonial population had doubled every generation or so, he took this to be a reasonable estimate of typical rates of human population growth. The importance of Malthus's famous calculation is certainly partly due to its immense usefulness in assuaging the conscience of the wealthy and undermining any social movements meant to address economic inequality—though it's not clear that this was his motivation for the original inquiry. Still, this kind of argument, appearing to rest on some underlying structure of nature, offers a handy justification for injustice. An example in our own time is the idea that "intelligence" is predominantly genetic, and thus attempts to affect it with more equitable distribution of wealth and opportunity are futile.

But here's the thing: Malthus was way off on his estimates of the rate of precivilized human population growth. Rather than doubling every generation, as Malthus supposed, archaeologists have shown that until the advent of agriculture, human population doubled roughly every *quarter million years*—not every twenty-five. Not once a generation, that is, but once every *ten thousand generations*.

No wonder there was such a gulf between the desperate, starving creatures Malthus imagined and the rather relaxed, healthy human beings actually encountered by explorers of his day, who spoke of people who were leading far richer lives than the vast majority of Londoners in 1800. They ate more nutritious food, worked less, slept more, and suffered less disease.

But observed reality is no match for comforting theory. Understandably, Malthus assumed that the injustices he saw around him were universally human and set out to explain them. But as often happens, his explanation ended up doing more to justify and perpetuate the conditions he witnessed than to accurately explain their origins or place them in the context of human possibility.

Even being off by a factor of ten thousand, the mistake proved useful. Malthus's greatest impact came through Charles Darwin and Alfred Russel Wallace, both of whom happened to be reading his essay when they independently came up with the theory of natural selection, on opposite sides of the world. In his autobiography, Darwin wrote that while reading Malthus's grim formula, "It at once struck me that under these circumstances favorable variations would tend to be preserved, and unfavorable ones to be destroyed. The result of this would be the formation of a new species." So the brilliant insight underlying natural selection—which has been called "the most powerful idea that ever occurred to a man"—was sparked by a colossal Malthusian miscalculation.

As for Thomas Hobbes, the poor guy was born to sing the blues. Even as a fetus, he shared the womb with terror when his mother went into premature labor upon learning that the Spanish Armada was just off the coast of England, about to attack. "My mother," Hobbes wrote, "gave birth to twins: myself and fear." Hobbes is long gone, but the fear lives on.

Things didn't get much easier later in life. *Leviathan*, the book in which he makes his famous arguments in support of an overpowering state to protect citizens from the ravages of nature and the savage impulses within, was written in Paris, where he was in hiding from British enemies who wanted him dead. He nearly succumbed to a six-month illness before completing the book. Once it was published, some of his fellow exiles in Paris decided they also wanted to kill him, so he fled back to England, begging for mercy from those he'd barely escaped a decade earlier. They let him stay, but prohibited his book, which was burned at Oxford.

Historian Mark Lilla describes the chaotic, frightening times in which Hobbes lived as "madness," where Christians "addled by apocalyptic dreams" persecuted other Christians with a "maniacal fury they had once reserved for Muslims, Jews and heretics." Hobbes, it seems, rendered his own very dark days slightly more palatable by imagining a prehistory so dire that even his world—bloody and chaotic as it was—seemed an improvement. We can't blame him for resorting to such comforting delusions, but nor must we repeat his mistake down the centuries.

It was 1651 when Thomas Hobbes described the paltry, pre-state world in which he imagined the uncivilized lived:

> In such condition there is no place for industry, because the fruit thereof is uncertain, and consequently, no culture of the earth, no navigation, nor the use of commodities that may be imported by sea, no commodious building, no instruments of moving and removing such things as require much force, no knowledge of the face of the earth, no account of time, no arts, no letters, no society, and which is worst of all, continual fear and danger of violent death, and the life of man, solitary, poor, nasty, brutish, and short (*Leviathan* XIII.9).

As it turns out, Hobbes's famous pronouncement is erroneous, false, untrue, specious, and wrong. But it was useful. Note how his depiction of precivilized life justifies the so-called White Man's Burden of bringing salvation to the primitives—even if you kill them in the process. After all, these are nasty, brutish, short-lived people, without any artistic or cultural sophistication, barely enduring their solitary, poor lives. Civilization can only be a vast improvement to these poor brutes!

It's impossible to overstate the political utility of the justifications Hobbes provided for the colonial enterprise that gave birth to the modern world. To argue for the essential humanity and dignity of people living outside the control of European authority (be it Church or State), as Bartolomé de Las Casas, Montaigne, and a few others tried to do, was to question the racial superiority of Europeans and the fundamental legitimacy of colonialism and the will of the Christian God as interpreted by men with vast armies at their disposal. The economic and political utility of these views—in centuries riddled with slavery, colonialism, and racism—can hardly be overstated. Less clear, perhaps, is why they still hold so much power today.

– The Functions of Fear –

Man is an animal suspended in a web of significance that he himself has spun.

—Max Weber

Richard Dawkins is one of the most famous scientists alive, and he is an enthusiastic teller of one of the darkest stories ever told. In *River Out of Eden*, Dawkins describes animal life as an operatic ordeal of starvation, misery, and pitiless indifference. "The total amount of suffering per year in the natural world is beyond all decent contemplation," he writes with trembling hand. "During the minute that it takes me to compose this sentence, thousands of animals are being eaten alive, many others are running for their lives, whimpering with fear, others are slowly being devoured from within by rasping parasites, thousands of all kinds are dying of starvation, thirst, and disease."

In Dawkins's telling, even the best of times only lead to the worst of times: "If there ever is a time of plenty," he says, "this very fact will automatically lead to an increase in the population until the *natural state of starvation and misery* is restored" (emphasis added). Let that sink in, if you dare. The "natural state" of living things is one of "starvation and misery." Very Old Testament!

During the minute that it took me to compose this sentence, how many people were sold on the dangerous, debilitating belief that the natural world is their lethal enemy and all that's keeping them from starvation, misery, and disease are the godlike wonders of civilization?

Pain and predation certainly exist, but so do the kindness of strangers, sunsets of indescribable beauty, rainbows on the domes of deep seashells, orgasms that—let's face it—feel far better than necessary, and mashed potatoes with garlic and butter. In any case, is the "total amount of suffering per year" meaningful? Wouldn't a far better metric be the *proportion* of one's life spent in agony versus quiet contemplation, blissful immersion, and simple satisfaction?

Dawkins is hardly alone in his dismal view of life outside the protective embrace of civilization. While such sentiments have been repeated for millennia, they may have reached their crescendo when the nineteenth-century philosopher Arthur Schopenhauer described the natural world as a "scene of tormented and agonized beings, who only continue to exist by devouring each other, in which, therefore, every ravenous beast is the living grave of thousands of others, and its self-maintenance is a chain of painful deaths."

It's a mistake to lose our sense of proportion, even when con-templating one's own demise—*especially* when contemplating one's own demise. It's true that we all must die eventually, but why be so dramatic about it? Contemplation of death is scary. I get it. But taken in context, it's a relatively brief event. In one of his last journal entries, Montaigne noted that as dying amounts to just a few bad moments at the end of life, it's really not worth worrying about. If it takes an hour to die, that would represent just one 700,000th of an average human lifetime. That's a pretty good ratio, when all is said and done. And if even one hour in 700,000 is too much for you, there are far quicker ways out—guaranteed painless—should you choose to take control of the process yourself or have a compassionate doctor.

And as for what comes next, what's to fear from that? As Mark Twain put it, "I do not fear death. I had been dead for billions and billions of years before I was born, and had not suffered the slightest inconvenience from it." But the NPP keeps warning that it's a jungle out there, and only the ramparts of civilization can protect us from "being devoured from within by rasping parasites" and the rest of nature, red in tooth and claw, waiting to pounce.

I recently came across an excellent example of this dark propaganda while watching one of those ubiquitous nature specials about great white sharks, "monsters of the deep," as the narrator kept calling them. The show opens with a cute seal happily frolicking in the waves for a few seconds before the ominous music starts to build, we catch a glimpse of a large shadow moving in the water, and a great white shark emerges from the depths and begins a slow-motion munching of the terrified, doomed seal (the narrator explains that the footage of the attack has been slowed down to one-fortieth normal speed, presumably to make every instant of horror easier to savor and harder to forget). We've all witnessed such scenes many times on TV, and it's hard to argue against the cruelty of nature when you're watching the flapping tail of a seal disappear down the throat of a sea monster, or an antelope twitching in the grip of a cold-eyed lioness. "Thank God I'm safe," we think, "sitting here on my sofa, with my Cheez Doodles and Big Gulp."

But I've had occasion to hang around some seals in my time, and they never struck me as particularly anxious animals. Every seal I've encountered was either snoozing on a warm rock or frolicking in the water with other seals. They looked happy, fit, and relaxed to me. Skeptical that a seal's lot in nature could be as bad as that slow-motion terror porn implied, I ran some numbers. It turns out that harbor seals live about thirty years. The gory death on that nature special took a few seconds in real time. So the ratio breaks down to roughly thirty years of hanging out with friends, eating fresh fish, and soaking up the sun followed by a sudden, unanticipated, nearly painless demise. Even if that particular seal died in her prime—at fifteen or twenty years of age—the ratio of pleasure to pain in her life was better than what most of us can expect.

Along with its indifference and occasional cruelty, nature has surprisingly compassionate qualities as well. One example is the euphoria-inducing compounds called endorphins that are released in mammals precisely when they're needed most. For obvious reasons, there are few firsthand accounts from people who have lived to describe the experience of being in the death grip of a predator, but the famous British explorer David Livingstone gave an unusually articulate account of having been attacked by a lion on one of his African expeditions:

> I heard a shout. Starting and looking half round, I saw the lion just in the act of springing upon me. I was on a little height; he caught my shoulder as he sprang and we both came to the ground below together. Growling horribly close to my ear, he shook me as a terrier does a rat. The shock produced a stupor similar to that which seems to be felt by a mouse after the first shake of a cat. It caused a sort of dreaminess in which there was no sense of pain nor feeling of terror, though quite conscious of all that was happening. . . . The peculiar state is probably produced in all animals killed by carnivora; and if so, is a merciful provision by our benevolent Creator for lessening the pain of death.

Although Dawkins respects Darwin above all thinkers, Darwin would certainly have found Dawkins's neo-Hobbesian terror of the natural world overwrought. At the end of a section of *Origin of Species* called "Struggle for Life Most Severe between Individuals and Varieties of the Same Species," Darwin wrote, "When we reflect on this struggle, we may console ourselves with the full belief that the war of nature is not incessant, that no fear is felt, that death is generally prompt, and that the vigorous, the healthy, and the happy survive and multiply."

George Orwell famously noted, "Who controls the past controls the future. And who controls the present controls the past." Indeed. And those who control the present have been misrepresenting the past for a long, long time. For centuries, we've heard the same scary stories. Back in 195 BC, Plautus declared that man is wolf to man (*homō hominī lupus est*). It's a short step from wolf-eat-wolf to Hobbes's five-part disdain for precivilized life, to data-defying proclamations about a bloody, desperate prehistory that never existed. The process of human self-domestication is fueled by vivid images of the ravenous beasts just outside the gates, waiting to devour anyone foolish enough to make a break for freedom. We are distracted from our immediate suffering by fairy tales about how terrible life used to be. And perhaps worst of all, many of us have been convinced that we carry the darkness within us, in our selfish genes. "It is simply human nature," we're told, "to rape and kill and enslave—and anyone who thinks otherwise is a foolish romantic."

This messaging not only offends our decency and dignity, it insults our intelligence. The depiction of human nature embedded in the NPP isn't science; it's a marketing campaign for the status quo. The politics of perpetual fear is corrosive to our well-being and our innate capacities for cooperation, community, and kindness. Fear of terrorists, fear of running out of money, fear of getting old, fear of strangers, fear of death, fear of sharks, fear of being hit by lightning, fear of fear itself. It keeps us quiet and complacent in our supposedly protective cages.

We're trapped in and by this distorting, demonizing view of human nature and the natural world, seen as the two faces of an enemy to be feared and conquered, rather than an ally to be honored and nourished. This pernicious nonsense has us divided against ourselves, each other, and the planet itself. We live under suspicion of our own and each other's natural impulses, ashamed to be animals, participating in the accelerating destruction of a natural world we've been taught is out to tear us limb from limb or gnaw away from inside. This is, all hyperbole aside, the deepest species-level psychopathology imaginable.

It would be hard to overstate how much the dual demonization of the natural world and of human nature has shaped modern sensibility. Politics, economics, foreign policy, criminal justice, our beliefs about the nobility of work, questions of how and whom we love, how we choose to give birth and opt to die—virtually *everything* we think and do rests on the conviction that the untamed and uncivilized are dangerous, merciless, evil, and "other."

To question the catechism of steady progress from primeval darkness into the light of civilization and modernity is to invite ridicule and scorn, largely because a ruthlessly competitive natural world is assumed to be the essential engine of both natural selection and capitalism. To be clear, it's true that Darwin argued that some individuals reproduce more successfully than others due to either natural or sexual selection. But differences in reproductive success don't require wildly unequal access to resources, endless misery, or early death for any of the creatures involved. Evolution isn't propelled by suffering. It works via differences in fertile offspring. One can be a total loser in terms of genetic legacy (as I am), having no descendants at all, and still live a long, happy life.

In his classic book *Walden*, Henry David Thoreau rebelled against the worship of "superior" men. Concerning the ancient

Egyptian pharaohs, he wrote, "As for the Pyramids, there is nothing to wonder at in them so much as the fact that so many men could be found degraded enough to spend their lives constructing a tomb for some ambitious booby, whom it would have been wiser and manlier to have drowned in the Nile, and then given his body to the dogs."

The NPP insists that we venerate the crooks, rapists, and pillagers credulous historians have repackaged as "founders," "conquerors," and "civilizers." We erect statues and consecrate tombs to commemorate their difference-making. But in fact, most of these monuments memorialize the dark deeds of unhinged lunatics driven by rampant ego and raving greed. "History," wrote Alexander Herzen, "is the autobiography of a madman," and in historical fact, most of the supposed "great men of history" were criminals on a rampage. We celebrate them because they "changed the world." But where's the evidence that they changed it for the better? Isn't it more parsimonious to conclude that the wake left by these ambitious boobies shaped civilization to reflect their own twisted values and ambitions? There is no logically sound reason to believe that the present is the predetermined destiny of the past. That's the twisted line of thinking used by those who proclaim, "I don't regret anything I've ever done, because if I changed anything, I wouldn't be me!"

"In that case, Mr. Manson, parole is denied."

In *Sex at Dawn*, Cacilda Jethá and I called the process of looking back from a perspective distorted by the present "Flintstonization." Take a look around and imagine a past based on what you see, just more primitive and rustic. After all, there's no arguing with the here and now, is there? Sure there is. George Bernard Shaw wrote that "patriotism is your conviction that your country is superior to all others because you were born in it." The same blind conviction contaminates our assumptions about the historical era we happen to have been born into. Let's call it "presentism." We're here now, so this is the best place to be!

But the mere fact that we happen to be here doesn't mean here is necessarily any better than worlds that have been trampled on and discarded en route. That this is the course that history happens to have taken doesn't mean it is the best possible outcome. To believe otherwise, one would have to believe in some kind of predestination, and argue that every toll paid along the way was worth it to get here: the Dark Ages, bubonic plague, millennia of slavery, unending war, uncountable genocides, disco—all of it. No doubt we've come a long way. But was it a long way up, a long way down, or just a long, long way?

– On Primitive Power –

Given a full chance to act in his own interest, nothing but expediency will restrain [a man] from brutalizing, from maiming, from murdering—his brother, his mate, his parent, or his child. Scratch an "altruist" and watch a "hypocrite" bleed.

—Michael Ghiselin, *The Economy of Nature and the Evolution of Sex*

In one of the most underlined passages of one of the most influential nonfiction books of the past century (*The Selfish Gene*), Richard Dawkins implores us to attempt to outwit what he sees as our deeply rooted, natural tendencies: "Let us try to teach generosity and altruism," he writes, "because we are born selfish. Let us understand what our own selfish genes are up to, because we may then have a chance to upset their design, something which no other species has ever aspired to." Stirring rhetoric, but if genes are inherently selfish, and we share roughly 98 percent of ours with chimps and bonobos—and a slightly lower percentage with other mammals—how could *our* altruism be a triumph over our genetically programmed nature, while the altruism documented among ants, dolphins, bees, herd animals, and many primates be somehow congruent with their genetic input? Although Dawkins is probably the greatest living popularizer of Darwin's work, their views of human nature appear to be worlds apart. Dawkins's human exceptionalism contrasts with Darwin's central conviction that "the difference in mind between man and the higher animals, great as it is, certainly is one of degree and not of kind."

It's tempting to think Dawkins meant it all metaphorically, but no. In *The Selfish Gene*, he is clear that the selfishness of humans is innate and encoded in our DNA. Dawkins views human beings as "survival machines—robot vehicles blindly programmed to preserve the selfish molecules known as genes," and the world of those genes is one of "savage competition, ruthless exploitation, and deceit." As goes the gene, so goes the man, because "this gene selfishness will usually give rise to selfishness in individual behaviour."

Ultimately, for Dawkins, "blindness to suffering is an inherent consequence of natural selection," but Darwin would have disagreed, as he believed that compassion and altruism conferred a clear evolutionary advantage on social animals. In his notebooks, Darwin observed, "Looking at Man, as a Naturalist would at any other mammiferous animal, it may be concluded that he has parental, conjugal and social instincts . . . these instincts consist of a feeling of love or benevolence to the object in question . . . such active sympathy that the individual forgets itself, and aids and defends and acts for others at his own expense." These ideas persisted in Darwin's thinking throughout his life and were expressed perhaps most eloquently in *The Descent of Man and*

Selection in Relation to Sex, published eleven years before his death, where he tells the story of a zookeeper he met: "Several years ago a keeper at the Zoological Gardens showed me some deep and scarcely healed wounds on the nape of his own neck, inflicted on him whilst kneeling on the floor, by a fierce baboon. The little American monkey who was a warm friend of this keeper, lived in the same compartment, and was dreadfully afraid of the great baboon. Nevertheless, as soon as he saw his friend in peril, he rushed to the rescue, and by screams and bites so distracted the baboon that the man was able to escape."

For Darwin, this cross-species selflessness was no aberration, but an expression of something fundamental in social species. "Many a civilized man who never before risked his life for another, but full of courage and sympathy, has disregarded the instinct of self-preservation and plunged at once into a torrent to save a drowning man, though a stranger. In this case man is impelled by the same instinctive motive, which made the heroic little American monkey, formerly described, save his keeper by attacking the great and dreadful baboon."

But for Dawkins, Steven Pinker, and other neo-Hobbesian thinkers, evolutionary advantage goes not to the altruist, but to the "selfish rebel." Dawkins lays this out clearly in *The Selfish Gene*:

> Even in the group of altruists, there will almost certainly be a dissenting minority who refuse to make any sacrifice. If there is just one selfish rebel, prepared to exploit the altruism of the rest, then he, by definition, is more likely than they are to survive and have children. Each of these children will tend to inherit his selfish traits. After several generations of this natural selection, the "altruistic group" will be over-run by selfish individuals, and will be indistinguishable from the selfish group.

This hypothetical scenario is foundational to the NPP and to the "rational self-interest" considered fundamental to capitalism. The mantra is repeated, virtually word for word, in any number of books and lectures. "Unless a group is genetically fixed and hermetically sealed," writes Pinker, "mutants or immigrants constantly infiltrate it. A selfish infiltrator would soon take over the group with its descendants, who are more numerous because they have reaped the advantages of others' sacrifices without making their own."

While this thought experiment seems to make sense in theory, it's reminiscent of the old saying "In theory, theory and reality are the same. But in reality, they are very different." Dawkins and Pinker are undoubtedly brilliant in their respective fields of biology and linguistics, but they appear to be unaware of the many measures foragers take to discourage "selfish infiltrators." These mechanisms are well known to anthropologists.

Christopher Boehm has studied politics and power in foragers for over four decades. When he combed through anthropological field reports on the 150 or so immediate return hunter-gatherer societies that have been studied by anthropologists, his meta-analysis revealed *the opposite* of what Pinker and Dawkins assume as their premise. Coding the reports for categories of social behavior such as aid to nonrelatives, group shaming, and the execution of social deviants, Boehm determined that *without exception* generosity and altruism are consistently favored toward relatives and nonrelatives alike. "Nomadic foragers are universally—and all but obsessively—concerned with being free from the authority of others," Boehm writes. "That is the basic thrust of their political ethos. . . . This egalitarian approach appears to be universal for foragers who live in small bands that remain nomadic, suggesting considerable antiquity for political egalitarianism."

Far from being admired as "clever opportunists," selfish individuals looking to exploit the generosity of other foragers are viewed as pitiful and potentially dangerous, likely to be nudged off the nearest cliff. Such an individual would be lucky to survive for long in a real-world foraging society, much less flourish. There's plenty of ferocity in the "fierce egalitarianism" of foragers.

Egalitarianism among foragers doesn't imply that there are no differences in ability or accomplishment, or that foragers don't have their own hierarchies. Rather, they are careful to assure that hierarchies of status and admiration don't interfere with equal opportunity and access to resources. Archaeologist Robert Kelly explains:

> The term *egalitarian* does not mean that all members have the same amount of goods, food, prestige, or authority. Egalitarian societies are . . . those in which everyone has equal access to food, to the technology needed to acquire resources, and to the paths leading to prestige. The critical element of egalitarianism, then, is *individual autonomy.* . . . Egalitarianism is not simply the absence of hierarchy. . . . The maintenance of an egalitarian society requires effort.

Boehm makes the counterintuitive argument that egalitarianism requires close attention to hierarchy. To maintain their egalitarian social groups, foragers constantly celebrate and reinforce their antihierarchical social codes. "If a stable egalitarian hierarchy is to be achieved," according to Boehm, "the basic flow of power in society must be *reversed* definitively" so that common people maintain the upper hand over those with ambitions that could upset the balance. We see this ancient democratic impulse at the heart of democratic ideals, and expressed in representative government. What is the message of "one person, one vote" and "all people are created equal" if not the articulation of this quintessentially antidominance disposition we've inherited from fiercely egalitarian ancestors? We contain multitudes, including impulses toward selfishness, but the impulse toward justice and individual autonomy extends millions of years into the substrate of the prehuman psyche, as primatologist Frans de Waal and others have demonstrated.

Opportunities for imbalances to develop are common, of course, so foragers employ tradition, humor, and ridicule to maintain social harmony. Among the !Kung San of Botswana, for example, credit for a kill—and, thus, the responsibility and honor of distributing the meat—goes to the owner of the first arrow to hit the animal. But since men trade arrows constantly in a system of reciprocal gift-giving called *hxaro*, the owner of the first arrow may not have even participated in the hunt, much less made the kill. The tradition of trading arrows effectively randomizes the distribution of status.

To keep the best hunters' egos underinflated, when the men go to help bring in the kill, they'll often complain about how miserable the animal is: "You mean to say you have dragged us all the way out here to make us cart home your pile of bones? Oh, if I had known it was this thin I wouldn't have come."

When a visiting anthropologist grew confused about why some men were being so dismissive of another's hunting success, a !Kung San man explained the situation: "When a young man kills much meat, he comes to think of himself as a chief or a big man, and he thinks of the rest of us as his servants or inferiors. We can't accept this. We refuse one who boasts, for someday his pride will make him kill somebody. So we always speak of his meat as worthless. In this way we cool his heart and make him gentle."

Half a world away, Eskimos in the Arctic cooled hearts in the same way, as Kent Flannery and Joyce Marcus explain in *The*

Creation of Inequality: "So crucial was food sharing that the Eskimos used ridicule to prevent hoarding and greed. . . . It was a truly egalitarian society in which the slightest attempt to hoard or put oneself above others was discouraged. A skilled hunter and good provider might be universally respected, but even he was expected to be generous and unassuming."

An associated factor not acknowledged by the "selfish infiltrator" narrative is that hunter-gatherers have been universally armed for at least half a million years. Weapons would have negated the effects of a stronger, larger male trying to coerce others or seize power. The constant presence of such weapons was "critical to the definitive reversing of hierarchies in prehistoric bands," according to Boehm. Beyond making it easier to dispose of any "selfish infiltrators" who didn't respond to gentler measures, Boehm believes the presence of lethal weapons may have even changed the anatomy and physiology of our species: "Weapons were in a position to transform political behavior by 500,000 years ago, a figure that provides fully 20,000 generations for weapons to affect the genetic selection of body size and build, display behavior, canine size, distribution of hair on the body, and possibly bipedal efficiency."

Foraging societies make short work of any loud-mouthed braggart trumpeting his supposed superiority over the rest. Even those considered to be admired leaders can't get away with self-aggrandizing behavior without losing status. And walking away was always an option, as foragers live in what anthropologists call "fission-fusion" social groups. Chimps and bonobos share the same social dynamic, suggesting it extends millions of years into our past. Groups come together and split apart as circumstances dictate: availability of food, seasonal weather changes, social tensions, and so on. In *The Origins of Human Society*, anthropologist Peter Bogucki explains that "Pleistocene bands were fluid associations of individuals whose affiliations were conditioned more by proximity and immediacy resulting in friendships among kin and non-kin than by any fixed set of biologically-determined relationships." In such "fluid" associations, in-group and out-group identity is ever changing and never truly fixed.

Further churning the composition of hunter-gatherer bands is the fact that humans appear to be a female-exogamous species, like chimps and bonobos. Upon reaching sexual maturity, the females typically leave the group they were born into and join another group—an observation supported by recent studies of mitochondrial DNA, in addition to decades of field reports.

Again, it's important to understand that natural selection isn't necessarily a battlefield; it's merely the result of subtle differences in reproductive success compounded over many generations. Nobody needs to be exploited or killed for this effect to accrue. Some individuals just need to have more surviving offspring than others. But the neo-Hobbesians continue to insist that our ancestors stumbled through a Malthusian hellscape, where every scrap of food provoked murderous conflict among the starving savages. This vision of our species' prehistoric past not only ignores the anthropological literature on contemporary foragers, it reeks of the colonialist insistence that we are "civilized" and they were "savages."

Despite its narrative power and ubiquity, this view of human prehistory is seen as outdated and inaccurate by most of the people who study actual hunter-gatherer life. Anthropologist Nurit Bird-David, for example, summarizes the scholarship on hunter-gatherer behavior as reflecting an *assumption of affluence* rather than the presumed scarcity central to the NPP: "Just as we analyze, even predict, Westerners' behaviour by presuming that they behave as if they did not have enough," she writes, "so we can analyze, even predict, hunter-gatherers' behaviour by presuming that they behave as if they had it made."

*   *   *

But what about well-documented evidence of human cruelty? What about war and concentration camps? In 1961, a psychologist named Stanley Milgram designed a study to investigate how people respond when authority figures command them to inflict pain on innocent strangers. Milgram reported that when told to do so, 65 percent of the study participants repeatedly administered what they believed to be increasingly painful electric shocks to a subject displaying obvious distress.

One could view Milgram's entire career as an attempt to understand and illuminate the depravities committed in the concentration camps of World War II. The first paragraph of his first published article contains a mention of the gas chambers. Coincident with his research first being published, Adolf Eichmann was on televised trial in Israel—the trial at which Hannah Arendt famously coined the phrase "the banality of evil" to describe what she saw unfolding.

In her book *Behind the Shock Machine: The Untold Story of the Notorious Milgram Psychology Experiments*, Gina Perry explains, "Milgram stressed the connection between Nazi functionaries like Eichmann and the subjects in his lab. His findings appeared to demonstrate that ordinary people would inflict pain on someone else simply because someone in authority told them to." Milgram's research appeared to have demonstrated the validity of neo-Hobbesian assumptions about human nature, and his research is still cited today as evidence of a deeply Hobbesian human nature. Each of us is a nasty brute at heart, held in check only by civilization. Milgram proved it.

But there's a problem. "This zombie-like, slavish obedience that Milgram described wasn't what he'd observed," according to Perry, who went back and inspected the original research notes. She points out that the commonly cited figure of 65 percent of people who followed the experimenters' orders and went to the maximum voltage on the shock machine was based on *just one* of twenty-four different variations of the study Milgram conducted, "each with a different script, actors and experimental set up." And that single variation involved just twenty-six subjects. In total, more than seven hundred people participated in the experiments, and their obedience rates varied enormously. In some scenarios, *none* of the subjects obeyed commands to shock the victim. In fact, Perry found that, overall, *most of the subjects had refused to inflict any pain at all*—quite the opposite of what millions of Psych 101 students have been led to believe.

Unsurprisingly, then, subsequent research has arrived at findings different from Milgram's. Molly Crockett and colleagues at Oxford University conducted a study in which subjects administered shocks to others or to themselves in return for money. Their results "contradict not just classical assumptions of human self-interest, but also more modern views of altruism," said Crockett. "Recent theories claim people value others' interests to some extent, but never more than their own. We have shown that when it comes to harm, most people put others before themselves. People would rather profit from their own pain than from someone else's."

Despite the many stories of wanton cruelty in war, even there, most people are deeply affected by the suffering of others. In *None of Us Were Like This Before*, Joshua Phillips interviewed American soldiers who'd abused prisoners in Iraq (often under orders). Phillips found that almost without exception, these men suffered intense guilt, PTSD, and substance abuse as soon as they left the war zone. Suicide was not uncommon. A survey carried out after the first Gulf War by David Marlowe, an anthropologist who later worked for the U.S. Department of Defense, showed that "combat veterans reported that killing an enemy soldier, or even witnessing one getting killed, was more distressing than being wounded themselves." But even worse, Marlowe found, was losing a friend. "In war after war, army after army, losing a buddy is considered the most devastating thing that can possibly happen. It is far more disturbing than experiencing mortal danger oneself and often serves as a trigger for psychological breakdown on the battlefield or later in life."

Clearly, most human beings—even after months of desensitization training and battlefield stress—cannot inflict or witness suffering without being traumatized themselves. This is a far cry from the creature described by Hobbes and his modern advocates.

If this neo-Hobbesian view of our species is so inaccurate, why is it so widespread? The popularity and persistence of scientific narratives often have more to do with how well they support dominant mythologies than with their scientific veracity. Milgram's findings were quickly and deeply woven into the cultural fabric because they support the NPP, not because they were true.

Human tendencies toward generosity and kindness are fundamental to human nature, not a thin layer of culturally enforced morality obscuring our innate selfishness like a cheap rug thrown over bloodstained floorboards. As primatologist Frans de Waal put it, "There never was a point at which we became social: descended from highly social ancestors, the monkeys and apes, we have been group-living forever." De Waal calls out the dark premise underlying so much theorizing in economics and evolutionary theory: the assumption that we are essentially selfish beings taught (or coerced) by a civilizing society to play nice with each other. On the contrary, de Waal believes that we are innately invested in cooperation at our deepest levels, and he agrees with Darwin that essential components of morality are subject to evolutionary processes. Thus, rather than "a human-made veneer," morality is "an integral part of our history as group-living animals, hence an extension of our primate social instincts."

Dawkins's contention that "we, alone on earth, can rebel against the tyranny of the selfish replicators" raises awkward questions concerning whether and where he sees humans fitting into the animal kingdom. By framing the human capacity for altruistic cooperation as a unique rebellion against genetic determinism, but the sociability of other group-living species as congruent with their genetics, Dawkins seems to be implying that humans are angelically exempt from the chromosomal constraints common to all other living things—a view that seems to place the world's best-known atheist in a pulpit.

Part II

APOCALYPSE ALWAYS
(THE NPP IN THE PRESENT)

Chapter 3

The Myth of the Savage Savage (Declaring War on Peace)

Man is the most vicious of all animals, and life is a series of battles ending in victory or defeat.

—Donald Trump

In accepting his Nobel Peace Prize, Barack Obama said, "War, in one form or another, appeared with the first man. At the dawn of history, its morality was not questioned; it was simply a fact, like drought or disease—the manner in which tribes and then civilizations sought power and settled their differences." When I heard these antiquated, discredited ideas articulated by such an intelligent, educated man, I was reminded of Mark Twain, who wondered "whether the world is being run by smart people who are putting us on, or by imbeciles who really mean it." A third option would be that the world is being run by smart people who have been misinformed by generations of scholars who were promulgating nonsense.

Archaeologist Raymond Dart, famous for having discovered the first fossil of a human ancestor in Africa in 1924, added some memorably gruesome visuals to the NPP when he described early humans as "carnivorous creatures, that seized living quarries by violence, battered them to death . . . slaking their ravenous thirst with the hot blood of victims and greedily devouring livid writhing flesh." Hope you saved room for dessert!

When not gorging on the hot blood and writhing flesh of their prey, our ancestors were presumably salivating over one another. A piece in the *New York Times* by science journalist Nicholas Wade assures readers that "warfare between pre-state societies was incessant, merciless and conducted with the general purpose, often achieved, of annihilating the opponent." In a book called *Demonic Males*, anthropologists Richard Wrangham and Dale Peterson characterized modern humans as "the dazed survivors of a continuous, 5-million-year habit of lethal aggression." No wonder Obama bought into this ubiquitous narrative.

The neo-Hobbesians present three primary types of evidence to support their view of endless prehistoric war:

1. Primatological data drawn mainly from chimpanzees, with whom we shared a common ancestor about five million years ago (hence, Wrangham and Peterson's "5-million-year habit of lethal aggression");
2. Anthropological data supposedly showing that contemporary hunter-gatherer people reflect the supposed brutality of our ancestors; and
3. Archaeological findings that they believe demonstrate persistent warfare extending back many millennia.

It is hard to say which leg of this stool is the wobbliest. I'll take them in order.

– Primate Evidence –

Pointing to chimpanzee group-level conflict to explain the origins of human war is a powerful rhetorical device. If war is an expression of something embedded so deeply in us that it goes back millions of years to before our ancestors diverged from the line leading to chimps (who sometimes engage in lethal group aggression), then war must be innate to our species.

But there are serious problems here. First, it's subtly, if deeply, misleading to describe chimps as "our closest primate cousin" without mentioning bonobos—our other, equally intimately related primate cousin. Bonobos tend to get mentioned in guarded whispers—if at all—in these sweeping declarations about the ancient primate roots of war. There are plenty of reasons easily embarrassed journalists might want to avoid talking about bonobos, such as their penchant for mutual masturbation, their unapologetic same-sex behavior and occasional incest, as well as the general bohemian shamelessness and leisure that pervade bonobo life. But the biggest inconvenience occasioned by bonobos may be the utter absence of lethal aggression among them. No war. No murder. No raping or pillaging. No infanticide. No support, in other words, for the primate origins of human war. Given that a common ancestor eventually evolved into humans, chimps, *and* bonobos, basic scientific and journalistic principles would seem to require that the bonobos' deeply antiwar ethos would get as much attention from serious authors as accounts of chimpanzee brutality. But that's not what happens.

In the *New York Times* article mentioned above ("When Chimpanzees Go on the Warpath," June 21, 2010), bonobos are mentioned just once, twelve paragraphs in. Wade describes bonobos as "the chimps' peaceful cousin" while chimps themselves are described as having a joint ancestor with humans—thus creating the impression that the human genome shares more with chimps than it does with the bonobos, which is simply false. Chimps and bonobos *both* descended from the same common ancestor that split from our line, so if one is our "cousin," both are. There is no scientifically justifiable reason to downplay or ignore the importance and relevance of bonobos in any discussion of "the primate origins" of any human characteristic or behavior. Bonobos are *at least* as relevant to human behavior as are chimps, if not more so, given the many traits we share with them and them alone.

The bonobo's absence is conspicuous not only in discussions of war. Look for the missing bonobo anytime a somber authority figure claims an ancient pedigree for human male violence of any sort. See if you can find the bonobo in this account of the origins of rape, from biological anthropologist Michael Ghiglieri's oft-cited book *The Dark Side of Man*:

> Men did not invent rape. Instead, they very likely inherited rape behavior from our ape ancestral lineage. Rape is a *standard* male reproductive strategy and likely has been one for millions of years. Male humans, chimpanzees, and orangutans *routinely* rape females. Wild gorillas violently abduct females to mate with them.

Ghiglieri mentions four great apes in support of his deep primate roots of rape thesis (humans, chimps, gorillas, orangutans), but there are five great apes. Bonobos are unmentioned, despite the fact that rape has *never* been witnessed in this species over decades of observation. Not in the wild. Not in zoos.

Bonobos aren't the only primates that undermine deterministic views of primordial pandemonium. Even famously quarrelsome macaques and baboons can learn to live peacefully if the right social pressures are brought to bear. Frans de Waal has written about an experiment in which rhesus monkeys (*Macaca mulatta*) and stumptail monkeys (*Macaca arctoides*) were housed together. The former are famously argumentative and bad tempered, while the latter are known for quickly reconciling after conflict. In the mixed group, the pushy, aggressive rhesus monkeys rapidly learned to chill out, and members of the two species were soon sleeping together in "large, mixed huddles."

Neuroscientist Robert Sapolsky witnessed a similar transition in a troop of baboons he was observing in Kenya. Contaminated meat from a nearby dump wiped out the most aggressive, high-ranking males in the troop—leaving less aggressive, lower-ranking males, who had no interest in harassing the females and young. Sapolsky feared these easygoing males would be powerless against the young males sure to infiltrate the troop in the next season. But upon his return to Kenya, he found new males in the troop who had adopted the easygoing approach rather than trying to overturn it. Clearly, there are serious problems with the primate origins of war theory.

– Anthropological and Archaeological Evidence –

Sadly, neo-Hobbesian discussions of the anthropological and archaeological literature can be just as limited as their forays into primatology. In his 2011 book, *The Better Angels of Our Nature: Why Violence Has Declined*, Steven Pinker argues that levels of violence and warfare are now far below where they were during prehistory, when "chronic raiding and feuding . . . characterized life in a state of nature." Without a single citation, Pinker lists a series of reasons foragers *must have* engaged in brutal warfare:

> Foraging peoples can invade to gain territory, such as hunting grounds, watering holes, the banks or mouths of rivers, and sources of valued minerals like flint, obsidian, salt, or ochre. They may raid livestock or caches of stored food. [Note: Livestock

and caches of stored food are two things foragers don't have, or they wouldn't be "foragers."] And very often they fight over women. Men may raid a neighboring village for the express purpose of kidnapping women, whom they gang-rape and distribute as wives.

Doug Fry and Patrik Söderberg, two anthropologists who specialize in the study of preagricultural societies, were surprised by these assertions. "Nowhere in the actual data [on nomadic foragers] are found instances of lethal raiding for trophies or coups, food caches, water holes, hunting grounds, river access, flint, obsidian, salt or ochre, or to gang rape or claim betrothed women." Fry and Söderberg conclude that there is "a meager degree of agreement between the actual nomadic forager data and Pinker's assertions about raiding," and that "nomadic foragers do not actually raid neighboring communities very much at all."

These distortions of how endemic lethal violence is in hunter-gatherer lives are not inconsequential. In fact, they form a necessary baseline for the central argument of Pinker's book, which is that "violence has declined over long stretches of time, and today we may be living in the most peaceable era in our species' existence." The archaeological evidence simply does not support this thesis. As Fry explains in *War, Peace, and Human Nature*, "The worldwide archaeological evidence shows that war was simply absent over the vast majority of human existence." Instead, the archaeological record is "clear and unambiguous" in showing that "war developed, despots arose, violence proliferated, slavery flourished, and the social position of women deteriorated" *after* our species shifted from foraging to living in large-scale agricultural settlements. Civilization has not reduced the ravages of human violence. On the contrary, civilization is the source of most organized human violence.

Reasonable people can disagree on what counts as homicide and what is "war," which foragers are most representative of how our ancestors lived, what kinds of skeletal evidence are relevant, and so on—but the interpretation I'm presenting here is not controversial among those who have studied foragers in any depth. Pinker often cites Dr. Robert Kelly, for example, who is anything but an outlier among archaeologists. He has authored more than one hundred articles, books, and reviews, including two of the most widely used university archaeology textbooks in the United States, and has served as department head at various universities and as the editor of the journal *American Antiquity*, the leading publication on archaeology in the United States. It would be hard to be more mainstream. In *The Foraging Spectrum*—a book whose title highlights Kelly's intention to accentuate the *variability* of forager societies—Kelly describes hunter-gatherers as living in "small, *peaceful*, nomadic bands, men and women with few possession[s] and who are equal in wealth, opportunity, and status" (emphasis added). Not mainstream enough for you? Pick up a copy of the *Cambridge Encyclopedia of Hunters and Gatherers*, and you'll read that nomadic foragers "have lived in relatively small groups, without centralized authority, standing armies, or bureaucratic systems." The authors stipulated, "The evidence indicates that they have lived together surprisingly well, solving their problems among themselves largely without recourse to authority figures and *without a particular propensity for violence. It was not the situation that Thomas Hobbes, the great seventeenth-century philosopher, described in a famous phrase as 'the war of all against all'*" (emphasis added).

So where do neo-Hobbesians find evidence to support their bloody claims? Pinker presents eight "prestate" societies he uses to establish a baseline for rates of death supposedly typical of our forager ancestors. I'm going to let Pinker slide on how representative eight contemporary societies could be of the general hunter-gatherer experience twenty thousand or more years ago (which is only fair, since I can be accused of the same sort of reasoning via selected example).

I'll mention but refrain from making a big deal of the fact that he has presented *horticultural societies* as being representative of *foragers*—as he did in earlier books and essays—without even addressing the discrepancy. After various scholars called him out on his highly problematic conflation of *hunter-gatherers* and *horticulturalists*, he stopped referring to his examples as "hunter-gatherers," switching to the slippery phrase "prestate societies" instead. Technically speaking, horticulturalists *are* prestate societies (if you accept the premise that foraging and horticultural societies are stages of development that lead inexorably to the state—a rather problematic, colonialistic assumption). But even accepting that premise for the sake of argument, horticultural societies are no more representative of foragers than teenagers are of infants, despite the fact that both are "preadults." Horticulturalists, *by definition*, have gardens, domesticated animals, and static settlements—all things that may be worth fighting over. These accumulated resources are absent in foraging groups, by definition.

More problems with Pinker's argument were exposed when Fry went back to the original ethnographic source material Pinker had used for his data on war deaths among foragers, including a 2009 article by Samuel Bowles, published in *Science*. Fry found that in two of the societies Pinker based his assessments on, the Aché of Paraguay and the Hiwi of Venezuela/Colombia, *"all of the so-called war deaths involved frontiersmen ranchers killing the indigenous people*, a tragic situation that has nothing to do with levels of warfare death in nomadic hunter-gatherers during the Pleistocene." Incensed at seeing the murder of native people by invading settlers used as evidence of the victims' supposedly innate warlike tendencies, Fry hammers the point: "To be absolutely clear, *the only so-called war deaths* reported are those where indigenous people were murdered or massacred by Venezuelans. *All* of these killings have been counted as so-called war deaths, as if they have relevance to estimating war-related deaths in the Pleistocene." One hopes this was simply a case of Pinker's not having read his source material closely enough to realize what he was doing, but to my knowledge, he has not offered any corrections or retractions.

In another essay in the book edited by Fry (*War, Peace, and Human Nature: The Convergence of Evolutionary and Cultural Views*), Brian Ferguson digs into Pinker's data and comes up with similarly disturbing results. Ferguson devotes an entire section to "Pinker's List" due to its crucial role in buttressing the argument that about 15 percent of the total population and a quarter or more of the adult men fell victim to the chronic warfare that supposedly plagued our prehistoric ancestors. "These numbers," writes Ferguson, "have become axiomatic." But "Pinker's list consists of cherry-picked cases with high casualties, clearly unrepresentative of prehistory in general." Ferguson goes into great detail showing the context Pinker has left out of his discussion, concluding that "the *total* archeological record of prehistoric populations . . . clearly demonstrates that war began sporadically out of warless condition, and can be seen, in varying trajectories in different areas, to develop over time as societies become larger, more sedentary, more complex, more bounded, more hierarchical." Ferguson concludes, "We are not hard-wired for war. We learn it."

Pinker's statistical analyses can be as misleading as his presentation of the data. Pinker cites the !Kung San of Botswana as an example of violent foragers often misunderstood by naïve observers: "The !Kung San . . . had been described by Elizabeth Marshall Thomas as 'the harmless people' in a book of that title," Pinker scoffed. "But as soon as anthropologists camped out long enough to accumulate data, they discovered that the !Kung San have a murder rate higher than that of American inner cities."

Those anthropologists must have camped out a good, long time. What Pinker fails to explain—or maybe understand—is that in a group of 150 people (a typical size for the !Kung San), a murder rate comparable to that of the deadliest American cities, of around twenty murders per one hundred thousand per year, would translate to one killing every thirty to forty years. Even if their statistical murder rate were *double* that of Baltimore or Detroit, there'd be, on average, one violent death per generation. Hardly the nasty, brutal existence Pinker paints.

Even Kelly, from whom Pinker drew many of his numbers, understands how misleading those numbers can be if not presented carefully. "The general tenor of daily social relations observed [among foragers] by the ethnographer," Kelly wrote, "can readily be a strongly positive one of friendship, camaraderie, and communal sharing that is very rarely disrupted by argument or physical fighting."

As long as even well-intentioned, deeply thoughtful, Nobel Peace Prize–winning political leaders are telling a story in which war is wrongly but confidently depicted as being as old as humanity itself, how can we move toward—or even envision—a world without war? The narrative claiming ancient origins of war functions both as an erroneous explanation of human nature and, tragically, as an impediment to the eradication of unnecessary savagery.

Chapter 4

The Irrational Optimist

The Right Honorable Viscount Ridley, hereditary member of the British House of Lords and former CEO of one of the largest banks in the United Kingdom, says not to worry. Be happy. Everything's great, and getting better all the time. In his unrelentingly enthusiastic paean to progress, *The Rational Optimist*, Matt Ridley argues that optimists tend to be judged unfairly, quoting the twentieth-century economist Friedrich Hayek: "Implicit confidence in the beneficence of progress has come to be regarded as the sign of a shallow mind." But Hayek's take on progress was far more nuanced than Ridley realizes. In the paragraph just after the sentence Ridley quotes, Hayek continues, "There never was much justification for the assertion that 'civilization has moved, is moving, and will move in a desirable direction,' nor was there any ground for regarding all change as necessary, or progress as certain and always beneficial."

Still, *The Rational Optimist* is certainly one of the most prominent examples of the "don't worry, be happy" genre. A brief foray into the arguments and rhetorical techniques Ridley employs will be illuminating, because the book offers a commonplace appraisal of modernity that is framed by a surprisingly uninformed depiction of precivilized human life and an apparent disregard for the costs of civilizational progress. The first page of the book, in fact, provides a clear indication of what's to come, featuring a brief epigraph and an illustration showing the growth of world GDP per capita over the past two thousand years. The graph doesn't address how wealth is defined, its relation to well-being, or how it is distributed. The epigraph is a quotation from Thomas Babington Macaulay: "On what principle is it that when we see nothing but improvement behind us, we are to expect nothing but deterioration before us?"

Any clear-eyed look over the historical shoulder will quickly reveal a great deal besides improvement behind us. In fact, *every complex civilization that has ever existed has collapsed into chaos and ruin* and there is no good reason to think the same won't happen to ours. In *Immoderate Greatness*, political scientist William Ophuls points to the inherent unsustainability of civilization, which he describes as "*Homo sapiens*'s bold attempt to rise above the natural state in which the species lived for almost all of its two hundred thousand years on Earth. Unfortunately," Ophuls continues, "by its very nature, this effort to become greater sets in motion a seemingly inexorable moral and practical progression from original vigor and virtue to terminal lethargy and decadence." Elsewhere in the book, Ophuls describes the sad predicament civilizations face: "A mature civilization is caught in an entropy trap from which escape is well nigh impossible. Because the available energy and resources can no longer maintain the existing level of complexity, the civilization begins to consume itself by borrowing from the future and feeding off the past, thereby preparing the way for an eventual implosion. . . . This is the tragedy of civilization: its very 'greatness'—its panoply of wealth and power—turns against it and brings it down."

"Nothing but improvement behind us?" Come on, now.

After a two-sentence nod to the fact that some people "still live in misery and dearth even worse than the worst experienced in the Stone Age," Ridley lays his cards on the table. "The vast majority of people are much better fed, much better sheltered, much better entertained, much better protected against disease and much more likely to live to old age than their ancestors have ever been." I don't know how Ridley proposes to measure how entertained our ancestors were, but all of his triumphant claims are, as we'll see, far more debatable than they appear.

– Mo Better Blues –

On the first page of his book, while rhapsodizing on the astounding success of *Homo sapiens*, Ridley writes, "By the middle of this century the human race will have expanded in ten thousand years from less than ten million to nearly ten billion people." But is this population explosion a reason for optimism or despair? When it comes to human population, bigger is only better in the early stages of agricultural development, when expanding societies are competing fiercely against each other for resources, trading routes, slaves, and so on. This belief that more people somehow translates into better, safer lives is a reflection of an outdated metric. In fact, the opposite is more likely to be true.

Consider the chicken. It was Charles Darwin who first suggested that a wild species in Southeast Asia known as red jungle fowl (*Gallus gallus*) is probably the ancestor of the modern chicken, a hunch confirmed by recent DNA testing. No one can say how many of these birds scrambled about in the underbrush before being domesticated, but it was surely a tiny fraction of the 50 *billion* chickens alive today.

But chickens raised for meat spend their lives in filthy sheds shared with tens of thousands of other birds. Selective breeding and the reckless use of growth hormones have resulted in animals that grow so quickly their legs often buckle under their own weight, and their internal organs are unable to function. Laying hens, on the other hand, typically live out their days in stacked wire cages, weathering unending storms of shit and piss raining down on them until their egg production drops, and they're sent off to slaughter. In what sense is *Gallus gallus* a successful species?

Not convinced by the chicken argument? The ten countries with the fastest-growing populations are Liberia, Burundi, Afghanistan, Western Sahara, East Timor, Niger, Eritrea, Uganda, the Democratic Republic of Congo, and Palestine. Which of these countries would you call a raging success? If ever there were a situation that demands that we value quality over quantity, the measure of what constitutes a good life must be it.

But what about our expanding wealth? Ridley claims, "The availability of almost anything a person could want or need has been going rapidly upwards for 200 years and erratically upwards for 10,000 years before that. Years of lifespan, mouthfuls of clean water, lungfuls of clean air, hours of privacy, means of travelling faster than you can run, ways of communicating farther than you can shout." And then he really hits his stride: "This generation of human beings has access to more calories, watts, lumen-hours, square feet, gigabytes, megahertz, light-years, nanometres, bushels per acre, miles per gallon, food miles, air miles, and of course dollars than any that went before. They have more Velcro, vaccines, vitamins, shoes, singers, soap operas, mango slicers, sexual partners, tennis rackets, guided missiles and anything else they could even imagine needing."

Far be it from me to criticize a man's passion for Velcro and mango slicers, but what? If we agree that quality of life is best measured in light-years, tennis rackets, and guided missiles, then yeah, I guess civilization takes the prize. But if you value community, personal autonomy, and a meaningful existence more than dollars, soap operas, and megahertz, you may come to a different conclusion. (And how anyone can argue with a straight face that the air and water are cleaner today than they were ten thousand years ago is beyond me.)

Ridley anticipates naysayers like me: "This should not need saying, but it does. There are people today who think life was better in the past. . . . This rose-tinted nostalgia, please note, is generally confined to the wealthy. It is easier to wax elegiac for the life of a peasant when you do not have to use a long-drop toilet."

But of course there *were* no peasants or long-drop toilets ten thousand years ago. This bait-and-switch technique is a well-established part of the neo-Hobbesian rhetoric. One minute you're trying to correct an inaccurate portrayal of Pleistocene foragers, and the next you're expected to defend medieval plumbing. Furthermore, Ridley's dismissal of his critics as being blinded by wealth is pretty rich, coming from a man who was raised in a castle, inherited an appointment to the House of Lords, and served as the CEO of one of Britain's largest banks until it collapsed under his leadership. "Blinded by wealth," you say?

– On the Health of Nations –

Ridley repeats the common belief that "the vast majority of people are . . . better protected against disease." But in fact, most of us are far more vulnerable to the most worrisome diseases now than people were in the Stone Age for the simple reason that, with very few exceptions, *the infectious diseases most deadly to human beings simply didn't exist in prehistory. They are by-products of civilization itself.* Before agriculture, humans didn't live with domesticated animals from which pathogens mutated into forms dangerous to our species. Only after agriculture did tuberculosis, cholera, smallpox, influenza, and the other well-known scourges of humanity emerge in population centers with densities sufficient to allow their spread once they'd mutated to human hosts.

The same can be said of many of the noninfectious diseases most lethal to our species. They are *caused by civilization, not alleviated by it*. The misalignment between our evolved physiology and the diet and lifestyle encouraged by Western civilization is behind many diseases of civilization. Coronary heart disease, obesity, hypertension, type 2 diabetes, many types of cancer, autoimmune disease, and osteoporosis—all are rare or absent among foragers.

Researchers have seen this process play out repeatedly in parts of the world experiencing so-called development. In an essay called "The Price of Progress," anthropologist John Bodley surveyed the health consequences typically suffered by people as their societies shift into civilization. First, as people enter the global economic system, they become vulnerable to diseases such as obesity and diabetes. Second, development disrupts pre-existing ecological balances, often resulting in higher rates of bacterial and parasitic diseases (for example, lots of standing water from construction projects can increase malaria). Third, when development fails (as it often does), it leaves once self-sufficient societies living in impoverished, filthy slums, subject to the many assaults on health associated with such conditions.

Bodley looked at the findings of an eight-member team of medical specialists, anthropologists, and nutritionists funded by the Medical Research Council of New Zealand and the World Health Organization who investigated the health of a genetically related population in the South Pacific at various points along a continuum of increasing involvement with the cash economy, the modern diet, and urbanization. After eight years of work, the team reported that they were "beginning to observe that the more an islander takes on the ways of the West, the more prone he is to succumb to our degenerative diseases. In fact," they concluded, "it does not seem too much to say our evidence now shows that the farther the Pacific natives move from the quiet, carefree life of their ancestors, the closer they come to gout, diabetes, atherosclerosis, obesity, and hypertension."

But what about dental health? Surely *that's* a vast improvement, right?

Not really.

When Buffalo Bill's Wild West show came to London in 1894, one thing that impressed Londoners about the Lakota Indians was their oral health. An article in the *Journal of the Royal Anthropological Institute* from that year reported that although half of them were in their forties or older, *none* of the ten Lakota had any cavities or missing teeth.

In the 1930s, an American dentist named Weston Price studied tribal people around the world to understand what conditions contributed to dental health. Price's travels took him to Alaska, the Canadian Yukon, Hudson Bay, Vancouver Island, Florida, the Andes, the Amazon, Samoa, Tahiti, New Zealand, Australia, New Caledonia, Fiji, the Torres Strait, East Africa, and the Nile. Wherever he went, Price found the same thing: If people were still eating their traditional diet, their teeth were perfect. But where they'd already begun the transition to a "modern" diet, cavities, missing teeth, and other abnormalities were common. The new diet brought with it reduced resistance to other diseases due to chronic oral infections that weakened the immune system, as well as "crowded, misplaced teeth, gum diseases, distortion of the face, and pinching of the nasal cavity." Like many earlier travelers to New Zealand, Price was struck by the robust health and fine features of the aboriginal Māori, but dismayed at the impact of the

Western diet, writing, "Their modernization was demonstrated not only by the high incidence of dental caries (cavities) but also by the fact that 90 percent of the adults and 100 percent of the children had abnormalities of the dental arches."

Skeletal remains of preagricultural people support Price's insights, showing that the cavities and gum diseases from which so many modern people suffer didn't arise until the grain-based diets of civilization and monoculture. Scientists analyzing skeletal remains found in modern-day Sudan, for example, concluded that less than 1 percent of the hunter-gatherers living in the area suffered from tooth decay. Once the local population took up agriculture, however, the rate quickly increased to around 20 percent. When our diet is in alignment with our species' evolved requirements, we don't suffer from tooth decay, and there is even evidence that our bodies can heal a decaying tooth.

The original meaning of the word "palliative" (dating to the fifteenth century) was care that "relieves the symptoms of a disease or condition without dealing with the underlying cause." Paleoanthropologist Daniel Lieberman and many other experts with a profound understanding of the environments in which our species evolved believe "evolutionary mismatches" are behind most of our health problems. If this is true, medical approaches that fail to acknowledge and address these essential conflicts are, strictly speaking, palliative rather than curative or preventive.

Consider the case of breast cancer, long known as "nuns' disease" because physicians had noticed that single, childless women were far more likely to get it than were married mothers. This observation led to large-scale studies documenting the correlation between higher rates of breast and uterine cancer and the number of menstrual cycles a woman experiences. The hormonal fluctuations a woman experiences with each cycle trigger cell division in her breasts, ovaries, and uterus. So far, so natural.

A girl in the industrialized world will, on average, begin menstruating at twelve years of age (or earlier in some populations) and continue until menopause, around forty years later. On average, she may get pregnant once or twice, and probably not until her late twenties or early thirties. If she breast-feeds at all, it will probably be for just a few months. Thus, over the course of her lifetime, she'll go through the ovulation/menstruation cycle from 350 to 400 times.

A typical forager female, on the other hand, begins ovulating at sixteen or seventeen, due to her much lower levels of body fat and no exposure to estrogenic contamination from plastics, growth hormones in livestock, and added sugar in her food. She's likely to become pregnant within a year or so of first ovulation (chastity being extremely uncommon in foraging societies) and will breast-feed each of her children for three to four years. Because lactational amenorrhea normally stops ovulation in breast-feeding women (and almost always in foragers with low body fat), such a woman—living in the physical and social world her body is adapted to—will menstruate only around eighty to one hundred times—about one-quarter as many times as a typical civilized woman. Since each of these menstrual cycles floods a woman's body with powerful hormones, it's not surprising that cancer rates in the affected tissues have exploded in modern times.

When Dr. James Larrick and his colleagues went to the Ecuadorian Amazon and examined the relatively isolated Waorani people, they found them to be among the world's healthiest human beings. They saw no sign of internal systemic failure syndromes such as heart disease, cancer, stroke, or diabetes. The Waorani had no internal parasites and showed no sign of previous exposure to polio, pneumonia, smallpox, chicken pox, typhus, typhoid, syphilis, tuberculosis, malaria, or serum hepatitis. Findings like these support the notion that heart and circulatory problems, cancer, stroke, and diabetes are largely, if not totally, due to misalignment between the world we've created and the one our bodies were expecting to inhabit.

– Food for Thought –

A major facet of the optimism trumpeted by Ridley and other proponents of the NPP is that we're "much better fed" today than people were in prehistoric times. This conclusion relies upon the neo-Hobbesian assumption that starvation was common until agriculture saved the day, which is approximately the opposite of true. The !Kung San foragers of the Kalahari desert, for example, eat an average of 2,140 calories per day, with ninety-three grams of protein. Because they rely on more than eighty wild plants, they are unlikely ever to face the starvation that strikes societies dependent upon just a few crops, which can and do fail. While foragers faced occasional food shortages, their mobility and varied diet allowed them to adapt to changing conditions in ways that modern populations simply cannot. Skeletal remains show that foragers faced occasional hunger but not extended starvation. Today, however, the United Nations Food and Agriculture Organization estimates that about 805 million people suffered chronic undernourishment in 2014. That's one in every nine people alive. To claim that "we're better fed" than foragers is to remove more than 800 million people from your understanding of what the word "we" means.

In the early 1960s, anthropologist James Woodburn took a team of medical researchers to assess the nutritional state of Hadza children in Tanzania. The researchers were surprised to find that their nutrition was excellent: no lack of protein, and none were below standard weights for their ages.

In his survey of the anthropological and archaeological research into these questions, *Health and the Rise of Civilization*, Mark Nathan Cohen extrapolates from findings like these to foragers generally, concluding that foragers "do surprisingly well if we compare them to the actual record of human history rather than to our romantic images of civilized progress." Evidence drawn from ethnographic descriptions of modern-day foragers, as well as the archaeological record, leads Cohen to conclude that *"the major trend in the quality and quantity of human diets has been downward"* (emphasis added). Indeed, says Cohen, "Even the poorest recorded hunter-gatherer group enjoys a caloric intake superior to that of impoverished contemporary urban populations."

Our species has gone from a situation in which the norm was for everyone to be well fed, with occasional, brief periods of hunger, to one in which almost 2 billion of us are obese or overweight (and many of those also malnourished), while more than 800 million are chronically hungry or literally starving to death. In what sense, exactly, is this "progress"?

And if you're put off by the thought of occasional, brief hunger, it turns out that a little hunger every now and then is surprisingly healthful. In fact, the *only proven technique for extending life span is caloric restriction*. More than seventy years ago biologists noticed that the reduction of calories by a third or more from what an animal would normally eat, if unlimited food were provided, extended the life span of fruit flies, rats, mice, dogs, and primates. The animals not only live longer, they are far healthier. Caloric restriction seems to have a protective effect against cancer, diabetes, and neuro-degenerative diseases. Roy Walford, a UCLA pathologist, found that mice fed about half of what they wanted lived about twice as long.

In an article published in the *American Journal of Clinical Nutrition*, Krista A. Varady and Marc K. Hellerstein summarize the many scientifically demonstrated benefits of caloric restriction in studies of both human and nonhuman animal subjects. A reduction of 15 to 40 percent from what the animal would eat with unrestricted access to food results in marked improvements in insulin sensitivity, cardiovascular health, and response to stress. These authors go on to point out additional benefits of occasionally eating less than we'd choose, including "increased average and maximal life span, reduced incidence of spontaneous and induced cancers, resistance of neurons to degeneration, lower rates of kidney disease, and prolongation of reproductive function."

Perhaps I've convinced you that the quality of foragers' lives is not as bad as you've been led to believe, but what about the quantity? I recall seeing a *New Yorker* cartoon that must have seemed clever to most readers, but made me want to punch a wall. Two "cavemen" in animal skins are sitting by a fire, and one of them says, "It's strange. We breathe clean air, drink clean water, eat only organic food . . . but still, we're dead by 30." Maybe our ancestors' lives weren't as nasty and brutish as we've been told, but they were short, right? Right?

– Longevity Lies and the Price of Paradise –

When I was a kid, dead baby jokes were all the rage at school for a year or two. I still remember a few of them. Although none was particularly funny, the "jokes" weren't about humor so much as touching a nerve—the jokey equivalent of a pestering tongue on a loose tooth. (*What's the best Christmas gift for a dead baby? A dead puppy.* Hilarious to a certain kind of ten-year-old.)

Some scholars have argued that the dead baby joke phenomenon began in the 1960s in the United States in response to the legalization of abortion and disturbing images coming back from

Vietnam. Few thoughts are as emotionally triggering to our species as the death of infants. Lest I be accused of romanticizing prehistory, let me be clear on this point: Foragers pay a very high price for their remarkable health, happiness, and personal freedom. And that price is exacted in a most precious currency: dead babies.

Among the aforementioned Hadza of Tanzania, for example, where researchers found amazingly healthy children, about one out of every five infants born dies in its first year, and 46 percent don't make it to the age of fifteen—rates that reflect the median values for a broad survey of foragers. There's nothing funny about that.

These high rates of childhood mortality are the key to clarifying confusion around human longevity. Imagine you and your family live on the south shore of a small island along with twenty other families. Everyone in your village is comfortable, but definitely middle class, with a family income of about $75,000 per year. But then Bill Gates decides to buy the uninhabited north half of the island and build a compound there. Gates has an annual income of about $11.5 billion, which means that the average family income on your island is now more than $500 million per year. How's it feel to be so rich all of a sudden? Not what you expected? Welcome to the illusory world of averages. Here's hoping your tax bill isn't based on the average income for the island.

When Mark Twain famously said, "There are lies, damned lies and statistics," he could have been talking about the statistics commonly cited to argue that human longevity has doubled or tripled thanks to civilization. "Life expectancy" has increased primarily because so many more infants and children now survive into adulthood. When infant mortality goes down, average life expectancy at birth goes up. When you include these early deaths in your calculations, "average life span" amounts to somewhere between thirty and forty years. *But a thirty- or forty-year-old human being has never been old.* Specialists from the fields of anthropology, medicine, evolutionary biology, and primatology all agree that our species' natural life span is around double that, and always has been.

In their comprehensive paper "Longevity Among Hunter-Gatherers: A Cross-Cultural Examination," anthropologists Michael Gurven and Hillard Kaplan discard slippery averages in favor of the modal age of death, which refers to "a peak in the distribution of deaths . . . the age at which most people experience sufficient physical decline such that if they do not die from one cause, they soon die from another." The modal age of death, in other words, is the age at which individuals in any given species are coming to the end of their natural lives. And the modal age of death for our species? Gurven and Kaplan couldn't be clearer: "The modal age of adult death is about seven decades, before which time humans remain vigorous producers, and after which senescence rapidly occurs and people die. We hypothesize that human bodies are designed to function well for about seven decades in the environment in which our species evolved."

For twenty-first-century Americans helped along with titanium hips, dialysis machines, and twenty-four-hour nursing home care, the expected age of death is about eighty-five, just a decade or so beyond when most hunter-gatherer adults die. A comprehensive review looking at the physiological data of primates goes further, finding that if a comparison group of monkeys and apes is used to make predictions based on anatomical similarities, a life span of ninety-one years is predicted for *Homo sapiens*. "The argument that human life span has not changed in 100,000 years can be considered substantially correct when the 'evolved' life span is considered," the authors note.

Misunderstanding and misrepresentation of the data on human longevity have caused generations of physicians and researchers to ignore overwhelming evidence that modern inactivity, stress levels, diets, and so on are pathogenic (disease causing). Many well-meaning physicians, for example, believe chronic back pain is the inevitable result of modern humans' living twice as long as our ancestors. Medical students are told that the human body is breaking down because it is being pressed into service for a far longer life span than it was designed for, like a 1958 Chevy still rumbling through the streets of Havana. Framed by this ubiquitous, erroneous understanding of human evolution, chronic pain, failing joints, cognitive disorders, and many other health issues arising after forty may appear to be signs of progress—not what they are: evidence of how modern life makes us sick.

In an interview with *NBC Nightly News*, for example, a biophysicist from UCSF explained, "It wasn't until two or three hundred years ago that we lived past age forty-five, so our spines really haven't evolved to the point where they can maintain this upright posture with these large gravity loads for the duration of our lives." Or consider this cascade of confusion from *Discover* magazine: "For the last century and a half, the average life span in wealthy countries has increased steadily, climbing from about 45 to more than 80 years. There is no good reason to think this increase will suddenly stop." Hold on. There is every reason to think that it will level off. As we run out of babies to save, infant mortality will stop declining, and this statistical sleight-of-hand will be revealed as the party trick it is. Misinformation about what's really been happening with average human life span has generated a slew of false clinical conclusions about how and when to treat patients, what sorts of preventive measures can and should be taken, and where to look for the true causes of poor health.

Misunderstandings about human longevity can also lead to disastrous policy decisions. When the second President Bush argued that Social Security should be privatized, part of his stated reasoning was that the system was biased against African Americans, because more of them died too soon to collect benefits. In a meeting with black leaders, Bush said that "African-American males die sooner than other males do, which means the system is inherently unfair to a certain group of people." But as economist Paul Krugman explained, "Mr. Bush's remarks on African-Americans perpetuated a crude misunderstanding about what life expectancy means. It's true that the current life expectancy for black males at birth is only 68.8 years—but that doesn't mean that a black man who has worked all his life can expect to die after collecting only a few years' worth of Social Security benefits. Blacks' low life expectancy is largely due to high death rates in childhood and young adulthood."

While a significantly greater percentage of infants died in prehistory than today, even *that* point isn't as unambiguous as it seems. First, many of those deaths were cases of what might be called "postnatal abortion" of children born in times of resource depletion (during a severe drought, for example) or with congenital deformities or other disabilities that would now be detected during prenatal testing, often resulting in an abortion. Such infants would not have survived long in a world where it was crucial to be mobile, vigorous, and sharp-eyed.

Infanticide is hardly a practice relegated to foragers, having been so widespread in Europe that foundling hospitals were opened to address the plight of infants being left to die by the side of the road. In the early 1800s, roughly a third of the babies born in Paris were left at the foundling hospital. For most of the infants, foundling hospitals offered little hope of survival. Of the 4,779 babies admitted to a hospital in Paris in 1818, for example, 2,370 died within three months. Other facilities had similarly dismal results. Half the infants admitted to the St. Petersburg hospital died in their first six weeks, and fewer than a third lived six years.

According to Chinese government records, about thirty-five thousand abortions are performed in that country every day. In China and India particularly, but not exclusively, healthy female fetuses are traditionally aborted because boys are preferred. My intention is not to debate the ethics of abortion, but to highlight the mathematical absurdity of including infant deaths in calculations of prehistoric life expectancy while excluding the many millions of abortions performed each year in estimations of contemporary life expectancy.

There's nothing funny about dying babies, but the exploding global population resulting from increased fertility, reduction in infant mortality, and religious resistance to birth control are no joke, either. Comparing high infant mortality (but low population growth and high quality of life) among foragers with the lower infant mortality rates of modern humans (but resulting exponential population growth and suffering of billions of impoverished people) can lead to difficult conclusions. Sarah Hrdy, one of the world's foremost authorities on hunter-gatherer parenting, believes that infant survival may be a far more nuanced issue than it seems, and that while many infants don't survive in hunter-gatherer societies, those who do can expect to be treated well, unlike many children in civilization:

> Those children who did survive back then were actually much better off in terms of the kind of nurturing environment that they experienced. Rates of child mortality were high, but there was no child abuse or emotional neglect. A child that has experienced the kind of emotional neglect it takes to produce the psychopathology of insecure attachment . . . simply would not have survived. Parents and other group members are very sensitive to anything that would threaten a child's survival. . . . Child abuse would not have been tolerated.

Nobody's ancestors—prehistoric or not—died in infancy, which means they could expect to live into their seventies or eighties. To keep spreading the idea that the human life span has doubled requires that one ignore facts about our species' innate longevity that have been comprehensively and repeatedly demonstrated. That's not science. It's advertising copy meant to sell the present.

Part III

REFLECTIONS IN AN ANCIENT MIRROR (BEING HUMAN)

Man is a creature that can get used to anything, and I think that is the best definition of him.

—Fyodor Dostoyevsky

"Character," they say, "is destiny." If so, the best way to predict the proper destiny of *Homo sapiens* is to understand the true character of our species. What kind of creature are we anyway? Hard to tell sometimes, as our sense of self, as a species, is distorted by the same neo-Hobbesian nonsense and self-aggrandizing civilizational propaganda that clouds our understanding of prehistory.

Human beings are adaptive creatures, but the fact that we *can* adapt to all kinds of horrible conditions doesn't mean we *should*. Like rats and cockroaches, *Homo sapiens sapiens* has found ways to survive and reproduce in conditions that would quickly have led most species to extinction. Right now, human beings are drilling into rock several miles from the nearest sunshine, floating far above the planet, scavenging city dumps, sleeping along railroad tracks, and trying to raise a family in the back seat of a car. But our capacity to adjust to such extraordinary conditions doesn't mean all our adapting comes easily or that these adaptive behaviors are necessarily "natural."

To ask "What is human nature?" is like asking "What's the natural state of H_2O?" So much depends on conditions. Liquid, solid, gas—temperature and pressure make all the difference. Similarly, human beings are capable of being egalitarian and selfish, violent and peaceful, cooperative and competitive. To a large extent, context is determinative. This is where many scientists stop talking about human nature: "We're so adaptive," they'll say. "We have a wide range of 'natural' behaviors." Which is true, as far as it goes. But it doesn't go far enough.

Human beings are complex and there is great variation among us in proclivities and behavior. Culture plays a powerful role in deciding what we consider to be "natural." What seems normal in one society can be considered inhuman in another: cannibalism, incest, infanticide, eating bunnies or puppies, and so on. Darwin noted how deeply our species can be shaped by cultural indoctrination, writing in *The Descent of Man*, "It is worthy of remark that a belief constantly inculcated during the early years of life, whilst the brain is impressible, appears to acquire almost the nature of an instinct; and the very essence of an instinct is that it is followed independently of reason."

As I've suggested, the particular "beliefs constantly inculcated" in our species for hundreds of thousands of years have not been selected for arbitrarily. They are the fruits of the remarkably consistent social and ecological conditions faced by our ancestors. Egalitarianism, cooperation, and the open sharing of resources were highly advantageous adaptations, just as Darwin predicted: "Those communities which included the greatest number of the most sympathetic members would flourish best, and rear the greatest number of offspring." The conditions of hunter-gatherer life—conditions shared by all our foraging ancestors—demanded an approach to other people that was egalitarian and fair. Because these demands formed an important part of our ancestors' social ecology for so long, they exerted a major, lasting impact on the development of our species and came to be deeply inscribed on our consciousness—"almost the nature of an instinct."

Santayana famously declared that "those who cannot remember the past are condemned to repeat it," but those who don't understand the distant past are condemned to live lives structured in ways that conflict with our deepest human appetites and tendencies. The modern world is the ultimate human zoo, designed, created, administrated, and occupied by humans. Tragically, the zoo we've designed for ourselves is a poor reflection of the world in which our species evolved, and is thus a profoundly unhealthy, unhappy place for too many of the human animals it contains. Human beings are *capable* of surviving in violent, confined contexts, but, like water, we grow stagnant and putrid when we cease to flow.

Civilization may be the greatest bait-and-switch that ever was. It convinces us to destroy what is free so an overpriced, inferior copy can be sold to us later—often financed with the money we've earned hastening the destruction of the free version. Contaminate streams, rivers, lakes, and aquifers with industrial waste, pesticide runoff, and fracking chemicals, and then sell us "pure spring water" (often just tap water) in plastic bottles that break down into microplastics that find their way to oceans, whales' stomachs, and our own bloodstreams. Work hard now so you can afford to relax later. We ignore friends and family while we struggle to get rich so someone will eventually love us. The voices of civilization fill us with manufactured yearnings and then sell us prepackaged dollops of transitory satisfaction that evaporate on the tongue.

Some throw up their hands and blame it all on human nature. But that's a mistake. It's not human nature that makes us engage in this blind destruction of our world and ourselves. For hundreds of thousands of years, human beings thrived on this planet without doing it in. No, this is not the nature of our species—it is the nature of civilization, an emergent social structure in which our species is presently trapped. To understand the roots of our seeming penchant for ecocide, we must understand that an animal's nature can only be expressed in relation to its environment, natural or contrived.

Until I stumbled upon a fascinating essay called "Die, Selfish Gene, Die" by science writer David Dobbs, I had no idea about the surprising relationship between grasshoppers and locusts. Grasshoppers, it turns out, are "elegant, modest, and well-mannered," by insect standards. Locusts, on the other hand, are trouble. The elegant grasshopper moves slowly on long legs and lives a quiet life, mostly in solitude. The locust "scurries hurriedly and hoggishly on short, crooked legs and joins hungrily with others to form biblical swarms that darken the sky and descend to chew the farmer's fields bare." Hard to imagine two creatures any more different. But here's the crazy part: Grasshopper and locust are in fact the same species. And not just the same species—one can morph into the other. They are the same animal. Got that? Same DNA, different critter. Not all species of grasshopper can become locusts, but all locusts were once grasshoppers.

In the most infamous species, *Schistocerca gregaria*, the desert locust of Africa, the Middle East, and Asia, the transformation is triggered by a familiar cascade of effects: surplus food leads to rapid population growth, the rains stop, food supplies dwindle, areas that are still fertile become overcrowded, resulting in higher population density, which kicks in epigenetic reactions, and once-elegant, relaxed grasshoppers become crazed, rapacious locusts. Wings and legs get smaller, coloring shifts—not over generations, but in the individual animals. Goodbye, chilled-out grasshoppers, hello, swarming, cannibalistic locusts. "Same genome, same individual, but," writes Dobbs, "quite a different beast."

I'm sure you can see where I'm going with this. While human DNA remains almost unchanged from preagricultural times, and we may not literally be classifiable as a separate species, the behavior of civilized human beings is every bit as distinct from that of hunter-gatherers as locusts are from grasshoppers. With the advent of agriculture, human population exploded, and we packed into overcrowded settlements for the first time in the existence of our species. Nearly everything about human life changed radically and rapidly: power dynamics, family structure, the status of women and children, the source and quality of food, our relation to other animals, our experience of disease and death, conflict with other expanding population centers—acquisition of land and property, what kinds of gods we worshipped and our relation to them . . . our place in the world and what sort of world it was. Move too slowly in this world and you're likely to fall victim to those behind you. The swarm may be the joy of the locust, but it is the ruination of the grasshopper. Grasshoppers don't *choose* swarming any more than our ancestors *chose* farming, or Brian Stevenson *chose* to rise above the morning fog.

In 1930, as Europe convulsed toward another world war, a scientist named Solly Zuckerman and his colleagues established a group of hamadryas baboons at a place called Monkey Hill, in the London Zoo. It wasn't long before all hell broke loose among the baboons, resulting in the deaths of 94 of the original 140 monkeys, including 14 infants. Zuckerman, a pioneer in using primate research to explore the underpinnings of human nature, attributed the massacre to a natural outbreak of "social discord," resulting from sexual competition among the males. He dismissed the possibility that there may have been something about their artificial environment that could have triggered the mayhem. "Behaviour is uniform," Zuckerman wrote. "The common belief that the new environment [of captivity] grossly distorts the expression of these relationships has no foundation in fact. The pattern of socio-sexual adjustments in captive colonies is identical with that observed among wild animals."

But Zuckerman was wrong. Subsequent research by ethologist Hans Kummer demonstrated that captive baboons are in fact *far* more aggressive than members of the same species living in the wild. Kummer found that females are nine times more aggressive, while captive males are more than *seventeen times* as aggressive, when living in cages.

Recall your own rage when trapped behind distracted idiots texting in traffic or wedged between smelly, snoring strangers in economy class while someone's demon spawn is kicking the back of your seat. Is your hostility an expression of human nature—or is it perhaps better understood as a minor facet of human nature magnified by the unnatural conditions you're trapped in?

In the 1970s and 1980s there were ubiquitous reports of laboratory rats repeatedly choosing drugs over food, again and again, until they died of starvation. As an ad by the Partnership for a Drug-Free America put it, "Only one drug is so addictive, nine out of ten laboratory rats will use it. And use it. And use it. Until dead. It's called cocaine. And it can do the same thing to you."

Bruce Alexander, a Canadian psychologist, decided to look more closely at these studies. Alexander and his colleagues ran a series of experiments centered on identical rats living in two different settings: One group lived in typical laboratory cages while the other group lived in a setting meant to replicate normal rat life as much as possible. The so-called Rat Park was two hundred times bigger than the cages, contained sixteen to twenty rats of both sexes, and plenty of food and toys. What Alexander and his colleagues discovered calls into question every behavioral study ever conducted on caged rats: The rats that were trapped alone in cages opted to get high as much as possible, but the rats with interesting lives (community, space, toys) tried the drugged water once or twice, and then stayed away from it. The rats with lives worth living had little interest in the escapism the drugs offered. Overall, they consumed less than a quarter of the drugged water the isolated rats did. None overdosed or ignored food until they starved. These studies strongly suggest that addiction may have more to do with traumatic experiences and environment than with the magical qualities of substances.

Since the behavior of animals in cages has at least as much to do with the cages as it does with any innate tendencies of the unfortunate creatures trapped within them, let's think carefully about the design of our own enclosures. Let us build human zoos that replicate our natural environment as closely as possible, allowing us to live lives that fit us like slippers, not high heels. When asked if there were lessons to be learned from the massacre at Monkey Hill, Frans de Waal said, "If you want to design a successful human society you need to know what kind of animal we are. Are we a social animal or a selfish animal? Do we respond better when we're solitary or living in a group? Do we like to live at night or in the daytime? You should know as much as you can about the human species if you have a hand in designing human society."

Despite the incessant marketing of modernity, many of us know in our bones that we are far from home, like grasshoppers in a world of locusts or baboons fighting it out on Monkey Hill. Amid all the bells and whistles, life here often feels cold, empty, isolating, meaningless, and barbaric. Our reality keeps drifting ever further from the ad copy. How many times have we heard about the next marvel that's "just around the corner"? Since 1970, the U.S. government has spent over $100 million on the "War on Cancer." But according to the National Center for Health Statistics, death rates from cancer dropped just 5 percent from 1950 to 2005 in the United States, and analysts from investment firms long ago noticed that *curing* disease is much less profitable than *managing* it. In 1970, it took eight hours to fly from New York to London. It still does, but now the seats are smaller. Armies of confused and angry young men rise up in rage against the unfulfilled promises of modernity in Africa, Europe, Asia, and the Americas. Even the things that *are* kind of amazing—space probes, smartphones, digital photography—quickly become expected parts of normal life and lose their luster, regaining importance only when they *don't* work.

Homo sapiens looks a lot like a species that has lost its way. The route leading to where we are only seems a path in retrospect. Looking back, it's clear we've been lurching from one thing to another with little understanding of what we were doing or where it all was leading. We have reached a pass that provides amazing perspective and potential. But still we're lost, with no fixed point from which to plot our course forward. If character is destiny, then perhaps our destiny can be found in a better understanding of our character.

Chapter 5

The Naturalistic Fallacy Fallacy

Generations have trod, have trod, have trod;
And all is seared with trade; bleared, smeared with toil;
And wears man's smudge and shares man's smell: the soil
Is bare now, nor can foot feel, being shod.

—Gerard Manley Hopkins, "God's Grandeur"

Is doesn't imply *ought*—but nor does *ought* imply *is*. The fact that something exists in nature doesn't mean it's necessarily healthy or wonderful. The natural world is replete with lethal snakes, poisonous berries, and infectious microbes. Nature is no place for carelessness, ignorance, or delusions of immortality. But the naturalistic fallacy—the belief that what's natural is always better— is only fallacious up to a point. While it's true that what exists in nature is not *necessarily* healthful, it's far more likely to comport with biological reality than something with no roots in the natural world. To deny the probability of an innate congruence of the natural is to adopt the *naturalistic fallacy fallacy*.

Consider shoe design. We can choose to ignore what "is" (the shape and function of the human foot) in arriving at what "ought" to be a great shoe design. And anyone who's spent hours in heels or pointy-toed business shoes will confirm that many shoe designers do just that. We end up with shoes that may look interesting, but walking or even standing in them for long makes for fashionable torture.

In *Born to Run*, his 2009 bestseller, Christopher McDougall explains how Nike convinced generations of joggers to ignore the evolved biomechanics of the human body to run in an unnatural, debilitating way that required the purchase of their expensive, utterly unnecessary products. Great for Nike's bottom line, but this departure from human biomechanics resulted in tens of thousands of injuries and incalculable costs to human health. McDougall quotes a financial columnist who thought Nike's plan was "brilliant." "[They] created a market for a product and then created the product itself. It's genius, the kind of stuff they study in business schools." You may think it's unfair to focus on Nike, but McDougall disagrees. "Blaming the running injury epidemic on big, bad Nike seems too easy," he writes, "but that's okay, because it's largely their fault."

Nike's "deftest move," McDougall writes, "was advocating a new style of running that was only possible in [its] new style of shoe [which] allowed people to run in a way no humans safely could before: by landing on their bony heels." In a profitable extension of the naturalistic fallacy, Nike rejected what *is* (the evolved design of human feet, knees, and spine) in proposing a new way that human beings *ought* to run. The results have been disastrous for everyone but shoe manufacturers. Daniel Lieberman has explained that "a lot of foot and knee injuries that are currently plaguing us are actually caused by people running with shoes that actually make our feet weak, cause us to over-pronate, give us knee problems." Before the advent of these shoes in 1972, Lieberman has noted, "People ran in very thin-soled shoes, had strong feet, and had much lower incidence of knee injuries."

Before 1972, humans had been running for a very long time. Our species has evolved many traits showing that our ancient ancestors were highly efficient long-distance runners. We ignore the inherited design of our bodies at our own peril. As Lieberman puts it, "Humans really are obligatorily required to do aerobic exercise in order to stay healthy, and I think that has deep roots in our evolutionary history. If there's any magic bullet to make human beings healthy, it's to run." But not the way they teach in business school.

McDougall calls it the "Nike Effect," but Nike is far from alone in following these steps to financial success. They just followed the process to spectacular wealth. We see the same process of replacing the cheap and natural with something worse in the "No Backyard Chickens, Industrialized Farming Effect," the "Unnecessary Cesarean Delivery on Friday so Your Doctor Can Golf on Saturday Effect," the "Growing Marijuana Is Illegal, Take These Toxic, Addictive, Expensive Pills Instead Effect," or the "Breast-feeding Is Disgusting, Use Formula Effect."

Replacing something natural, healthful, and free with something that promises a lot but delivers trouble is as old as agriculture, as old as civilization itself, in fact. It's what keeps the gears of commerce spinning. As early as 1930, American business consultants openly, excitedly explained that "advertising helps to keep the masses dissatisfied with their mode of life, discontented with the ugly things around them. Satisfied customers are not as profitable as discontented ones."

Just as the shape and function of the human foot are vital considerations in the design of a decent pair of shoes, an accurate understanding of hunter-gatherer experience is essential to living a satisfying, healthy life today. The stability and continuity of the foraging life over hundreds of thousands of years is both evidence of its utility and the original source of our humanity. As Nassim Nicholas Taleb points out in *Antifragile*, "Time is the best test of fragility—it encompasses high doses of disorder—and nature is the only system that has been stamped 'robust' by time."

We are certainly capable of ignoring the subtle dictates of our evolved, time-tested nature, but we pay a high price for doing so. I can defy my body's need to move in favor of sitting here writing this book, but my risk of heart disease, obesity, diabetes, stress-related mental illness, and other ills will increase as a result. We can deny our naturally gregarious sexual appetites, but those distorted energies will find expression in frustrations, resentments, and psychopathologies of various kinds. We can survive on five or six hours of sleep, but we'll suffer reduced cognitive function, depressed immune response, and a host of other psychophysiological problems. So to those who proclaim our ability to override our evolved nature, I say, "Well, only up to a point."

Chapter 6

Born to Be Wild

The greatest terror a child can have is that he is not loved, and rejection is the hell he fears. I think everyone in the world to a large or small extent has felt rejection. And with rejection comes anger, and with anger some kind of crime in revenge for the rejection, and with the crime guilt—and there is the story of mankind.

—John Steinbeck, *East of Eden*

Cleanliness may be next to godliness, but when it comes to delivering babies, messier is better. In a study of children born within a few days of one another in the same hospital in Brazil, those delivered by C-section were found to be missing the "starter biome" that babies delivered vaginally got naturally from their mothers. These too-clean infants ended up colonized by bacteria from other far less beneficial sources, ranging from the doctors and nurses in the room to the lampshades and bedsheets. "The founding populations of microbes found on C-section infants are not those selected by hundreds of thousands of years of human evolution," explained Dr. Martin Blaser, whose wife and colleague, Maria Gloria Dominguez-Bello, first conducted this research. The first microbes to colonize an infant's skin, eyes, and digestive system may have lifelong effects on the child's health, and Dominguez-Bello believes these early microbial interactions may be crucial in configuring an immune system that properly distinguishes "self" from "nonself."

Dr. Dominguez-Bello has applied these insights to follow-up studies in Puerto Rico and at NYU, where babies delivered by C-section receive swabs of their mother's vaginal fluids on their lips, face, chest, arms, legs, back, genitals, and anal region. While the bacterial colonies of these babies were still not as rich as those of infants who'd been born vaginally, they were far more closely aligned with their mothers' than were those of C-section babies who hadn't received any vaginal fluids. Studies like these are still in the early stages, but their importance is hard to overstate. Babies delivered surgically appear to be at higher risk of developing various immune and metabolic disorders, including type 1 diabetes, allergies, asthma, and obesity.

Fatherhood can be messy, too. The reality of how foragers often approach fatherhood undermines neo-Hobbesian assumptions concerning the universality of individualism, male dominance of women, "ownership" of children, and so on. What anthropologists call "partible paternity" has been widely documented among disparate groups in South America, including the Aché of Paraguay, the Mehinaku, Kaingang, Araweté, Canela, and Curipaco people of Brazil, the Matis of Peru, and the Yanomami and Bari of Venezuela. The idea, common to at least six different linguistic groups, is that a fetus is literally made of all the sperm a woman receives from various men over a period of ten months or so before birth. A woman who hopes to have a child who is strong, smart, funny, and a good hunter will make a point of having plenty of sex with men who are strong, smart, funny, and handy with a blowgun. When the child is born, men who have "contributed" to the baby will all consider themselves to be members of Team Dad. One might assume that babies of muddled paternity would be at greater risk of abandonment and early death, but the opposite is true. Steven Beckerman's research on Bari children revealed that those with a single father had a 64 percent chance of surviving until age fifteen, while the presence of an additional "father" increased those odds to 80 percent. Similar advantages have been observed in other societies that practice partible paternity.

In egalitarian societies in which sharing was the central organizing principle and private property was nonexistent, there would have been little reason for concern over paternity. The presumptive nuclear family is an artifact of civilization, where unwed mothers have for centuries been abandoned at best, shamed and even murdered at worst. When male-female relations were reframed in newly agricultural societies, the mutual respect and autonomy characteristic of foragers were replaced by something closer to a master-slave dynamic. This tragic and lasting collapse of human dignity was largely driven by a demand for paternity certainty among newly possessive males who now wanted to know who was going to inherit their accumulated wealth.

An additional reason to doubt the neo-Hobbesian parenting narrative is that many men in hunter-gatherer societies report feeling *least* attracted to women precisely when the women's need for provisioning would be highest—at extreme pregnancy or just after giving birth. As Sarah Hrdy explains: "It is telling, for example, that among people like the Aché, hunters lose interest in their wives right after birth—at just the time when a woman most needs to be provisioned. . . . Nothing about this pattern conforms to predictions generated by the model that either women's sexuality (willingness to engage in sex) or their sexual attractiveness evolved to insure male provisioning after birth."

Hrdy's point is not that men evolved to be unreliable cads—at least, that's not her *only* point. Yes, the standard, nuclear-family-celebrating story of one husband providing for his mate and their children is a far cry from what most anthropologists have observed. And yes, pregnant women and new mothers *are* highly vulnerable and in need of significant assistance. So who helps out? In some cultures, as noted, a woman may elect to recruit several fathers for each of her pregnancies, but in *all* foraging societies, such women could rely on the assistance of pretty much everyone: "A Pleistocene mother . . . was likely to be a mother embedded in a network of supportive social relationships," writes Hrdy. "Without such support, few mothers, and even fewer infants, were likely to survive."

Mainstream narratives of human sexual evolution imagine the role of "provider" having been privatized to an individual man, like some kind of 1950s suburban housing development, but, in fact, provisioning was provided by the band in general. And if, by some unfortunate circumstance, it wasn't, the mother and infant were unlikely to survive long. Tragic as such a situation would be, on another level, it meant that only happy, loved, wanted children grew up to carry forward the mutually supportive values of that society.

In an egalitarian band, a child benefits from communal support and love of many adults. In such groups, it's customary for any nearby adult or juvenile to pick up a fussy child. To take the Efé of the Congo as an example (and a representative example, at that), anthropologist Melvin Konner has shown that individuals other than the mother account for almost 40 percent of physical contact with a baby at three weeks of age, and 60 percent at eighteen weeks. He found that "infants [were] passed from hand to hand 3.7, 5.6, and 8.3 times per hour at 3, 7, and 18 weeks, respectively." Each infant was taken care of by anywhere from 5 to 24 different people (mean=14.2). And "attempts to comfort an infant within ten seconds of fussing occurred 85 percent of the time in the first seven weeks and 75 percent at 18 weeks."

It's the same story among the Aka, a hunter-gatherer group also in the Congo region of Central Africa. Anthropologists report that "infants are held almost constantly, they have skin-to-skin contact most of the day . . . and they are nursed on demand and attended to immediately if they fuss or cry." This is how human infants are meant to be treated. In *The Continuum Concept*, Jean Liedloff explains that this precognitive sense of being welcomed and loved is an essential element in the experience of our species:

> The feeling appropriate to an infant in arms is his feeling of rightness, or essential goodness. The only positive identity he can know, being the animal he is, is based on the premise that he is right, good, and welcome. . . . There is no other viable way for a human being to feel about himself; all other kinds of feeling are unusable as a foundation for well-being. *Rightness is the basic feeling about self that is appropriate to the individuals of our species.* . . . A person without this sense often feels there is an empty space where he ought to be. [emphasis in original]

That's where we come from, but it's a far cry from where we are now. Children who are unloved and unwanted survive in our times—which seems to be a good thing. But Hrdy argues that technological advances "to some extent decoupled (infant survival) from continuous contact with mothers and other care-givers," and this decoupling has resulted in thousands or millions of men and women maturing in postagricultural societies who would never have survived in the ancestral environments. Their survival has resulted in a world full of "ragged bands of street urchins" and "orphans in refugee camps" who have survived "all manner of neglect." And even the lucky ones may not be so lucky, as even wealthy suburban kids survive parenting that an Efé or a !Kung mother would see as negligent. "Never before in the history of humankind," Hrdy notes, "have so many infants deprived of social contact and continuous proximity to caretakers survived so well to reproduce themselves so successfully." Take a moment to ponder what ambitions and hungers these children carry into adulthood, and how they play out in politics, business, and crime. Is this "progress," or something else?

The infant mind is determined by these early interactions, and they set the range of experiences available later in life. This makes perfect evolutionary sense, as the first job of any creature is to get a sense of the environment it will be facing. For a social mammal like *Homo sapiens*, this means sussing out what can be expected from other people. Are they loving and kind? Can I trust these people? A child who breast-feeds for as long as she wants (typically four or five years in foraging societies), is always welcome on someone's hip, and feels the quick and ready affection of dozens of adults is likely to answer "yes" to these questions. A child who feels abandoned and alone, crying herself to sleep in a terrifying, dark room every night, with only sporadic physical touch from just one or two adults is likely to conclude that she can't really trust the adults in her life. And these answers will determine the nature of the child's life. Liedloff explains: "When his later experiences do not correspond in character to the ones that conditioned him, he tends to influence them to acquire that character, for better or worse. If he is accustomed to loneliness, he will unconsciously arrange his affairs to assure him a similar level of loneliness. Attempts on his own part, or of circumstances, to make him very much more or less lonely than is customary to him will be resisted by his tendency to stability."

We might say that these first infant experiences form the substance of his own origin story. And like all origin stories, this one ultimately creates the world in which it's told. Looking at parenting practices, it's easy to see why the personal origin story of the typical forager would be something like: "I am loved and welcomed and respected by these people whom I can trust, and I will live my life in a world that can be dangerous, but is normally generous and glorious." Sadly, it's also easy to see why the personal origin story of so many of us born after the agricultural revolution would assume a decidedly Hobbesian tone: "I am confused and alone. They abandon me in the darkness for endless hours, ignoring my cries of terror. There must be something wrong with me. They don't want me. I'm helpless and alone. I don't know who I can trust. I am cursed. This life will be a struggle for survival." When Thomas Hobbes wrote that his mother "gave birth to twins: myself and fear," he could have been referring to a great many mothers in civilized societies.

Psychologist Darcia Narvaez, who studies the moral development of children, has identified six characteristics of child rearing that she believes to be essentially human:

- Plenty of positive touch in the form of carrying, cuddling, and holding—but no hitting or spanking;
- Quick response to a baby's cries: "Warm, responsive caregiving [that] keeps the infant's brain calm in the years it is forming its personality and response to the world";
- Anywhere from two to five years of breast-feeding;
- Multiple adult caregivers who love the child;
- Lots of free play with multiage playmates; and
- Natural childbirth, which provides the mother with hormonal surges that appear to be protective against postpartum depression and provides the child with immediate and lasting immunological advantages, as explained earlier.

Narvaez believes the fact that today's university students are 40 percent less empathic than they were just thirty years ago may be due to increasingly maladaptive parenting. Recent increases in ADHD, aggressive behavior, anxiety, and childhood depression suggest something is seriously wrong—and getting worse. "The way we raise our children today in the US is increasingly depriving them of the practices that lead to well being and a moral sense," she warns.

The damage done to children by this culturally sanctioned emotional abuse is difficult to treat, at best. Just as a fish born into frigid water will take that frigidity as a baseline temperature, a child born into a cold or cozy emotional space will assume that to be the normal condition and extrapolate from there. Hrdy sees the same process playing out, chillingly, at the species level: "If the human capacity for compassion develops only under certain circumstances, and if an increasing proportion of the species is surviving to breeding age without developing these capacities, it won't make any difference how beneficial compassion was among our ancestors. . . . No matter what the dividends might have been in terms of high levels of interpersonal cooperation, natural selection cannot continue to favor a genetic potential that is not expressed."

By and large, parents try to do what the experts advise, and few have the temerity to question whatever the reigning medical guidance may be. In the early twentieth century, doctors assured mothers that their infants would benefit from being placed in sterile isolation chambers, where thousands of them died from the simple lack of physical contact with another living being. This was *not* the fault of the desperate, well-intentioned mothers who stood by helplessly while their sons and daughters drifted away. One could even argue that it wasn't the fault of the doctors who were, after all, acting in what they believed was the best interest of the children. We're given similar advice concerning formula over breast milk, Cesarean delivery over vaginal, making the child sleep alone over being in bed with parents. It's easy to be wrong when one is arguing against natural processes refined over thousands of generations.

We've all seen them and possibly been them: the stressed-out, struggling parents investing their last dime on ungrateful little brats who vengefully and loudly announce that they wish they'd never been born at all. Miserable parents, miserable kids. Strange way for the planet's "most successful" species to reproduce. And it seems that this woeful impression isn't based on just an occasional bad day, at least in the United States. Researchers found that the so-called happiness tax paid by American parents is the highest in the developed world. Americans without children were far happier than those with children—a gap significantly larger than that found in the United Kingdom, Australia, or the other twenty-two cultures studied. Apparently, American parents are miserable because they find themselves on the wrong side of the vaunted individualism so central to their national self-image. "The negative effects of parenthood on happiness," concluded the researchers, "were *entirely* explained by the presence or absence of social policies [that support parents]." In Denmark, Sweden, and other countries with social policies that helped parents combine child care with their work responsibilities, there was no happiness gap at all. *My* kids. *My* family. *My* problem.

And the kids have it just as bad. The authors of the Duke Child and Youth Well-Being Index recently reported that "children's health has sunk to its lowest point in the 30-year history of the Index." Nearly 8 million American kids suffer from mental disorders, with *prescriptions for psychotropic drugs for kids up 49 percent just between 2000 and 2003*. A not unrelated finding is that between 1997 and 2003, there was a drop of 50 percent in the proportion of children aged nine to twelve who reported spending time hiking, walking, gardening, and so on, according to research by Sandra Hofferth, a research professor at the Maryland Population Research Center and expert on how children spend their time. In a similar finding, researchers in Scotland clipped small devices to the waistbands of seventy-eight three-year-olds for a week. They concluded that the toddlers were physically active only about twenty minutes per day.

And American children are far more likely to be physically abused than children in other developed countries. Between 1994 and 2004, approximately twenty thousand American children were killed by family members in their own homes. That's four times the number of American soldiers who died in Iraq and Afghanistan during the same period, according to data compiled by Michael Petit, president of Every Child Matters. The child maltreatment death rate in the United States is triple Canada's and eleven times that of Italy.

But Petit is not blaming parents, who are struggling with a social system that is flawed at best, pathological at worst. When you receive no significant support from your society and have to work two jobs just to pay for the day care that allows you to go to work, nobody can blame you for putting your kids in front of the TV, feeding them whatever you can afford, and not wanting to spend the night comforting them when they're restless. Many progressive European societies have policies that replicate hunter-gatherer parenting values by assuring community support for parents via generous maternity and paternity leave, subsidized medical and child care, and free education.

Parents in the United States and other societies less aligned with the deeply human, communal values are struggling—not because they are bad parents, but because their culture places wildly unrealistic demands and expectations on them, abandoning them when they most need help. The United States has one of the highest relative child poverty rates in the developed world, according to UNICEF, finding that children's material well-being was highest in the Netherlands and in the four Nordic countries and lowest in Latvia, Lithuania, Romania, and the United States.

Anthropologists agree that "very close mother-infant contact, late weaning, and indulgent responsiveness to infant crying were highly characteristic of the hunting-gathering groups." Tragically, the same studies show that, whether measured by body contact, sleeping distance, response to crying, or weaning age, mother-infant contact and maternal indulgence of infants were lower in the United States than in the 176 less "advanced" cultures included in the study. But again, I want to stress that *this is not a failure of parenting but a failure of civilization.* Few mothers wouldn't choose to spend quality time with their kids if they could, but not alone and not after working all day. Human beings are "cooperative breeders," to use Hrdy's terminology. It is our nature to raise our children communally, but the modern world too often blocks that option, fracturing opportunities to raise children in ways they and their parents are evolved to expect.

Violent, aggressive societies have a natural tendency to replicate themselves. When developmental neuropsychologist James Prescott conducted a meta-analysis on tribal cultures, he found that it was possible to predict with 80 percent accuracy the peaceful or homicidal violence of forty-nine tribal cultures from a single measure of bonding between the mother and her child. The peaceful or violent nature of the other ten cultures studied could be predicted by whether youth sexual expression was supported or punished. "In short," Prescott wrote, "these two measures of affectional bonding . . . could predict with 100% accuracy the peaceful or violent nature of these 49 tribal cultures distributed throughout the world." Other researchers have found statistically significant correlations between low mother-infant contact and higher "frequency of drunkenness" later in life, more reported violent behavior, greater frequency of suicide, depression, and behavioral problems.

Richard Louv, author of *Last Child in the Woods: Saving Our*

Children from Nature-Deficit Disorder, has attributed much of this mess to recent cultural shifts in the American experience of nature over the past century, "from direct utilitarianism to romantic attachment to electronic detachment." Louv reported that "each hour of TV watched per day by preschoolers increases by 10 percent the likelihood that they will develop concentration problems and other symptoms of attention-deficit disorders by age seven." The mismatch between the human animal and the demands of society is profound and tragic. "The real disorder," Louv wrote, "is less in the child than it is in the imposed, artificial environment. . . . To take nature and natural play away from children may be tantamount to withholding oxygen."

Chapter 7

Raising Hell

And surely there is in all children . . . a stubbornness, and stout-
ness of mind arising from natural pride, which must, in the first
place, be broken and beaten down; that so the foundation of
their education being laid in humility and tractableness, other
virtues may, in their time, be built thereon.

—John Robinson (pastor of the Pilgrim colony)

If modern children are being denied the oxygen of experience,
it may be because well-intentioned parents have been led to
believe that the world is too dangerous a place for unsupervised
play. We wipe our little miracles with antibacterial solutions,
lead them around on protective leashes, encase their heads in
helmets, and isolate them from strangers. But as is so often the
case, while we're busily defending against largely imaginary dangers, we're creating real problems. Despite all the media reports
of abducted, murdered innocents, kids face less danger today
than when you and I were catching frogs in the woods or playing
under the streetlamps. The overall child mortality rate in the
United States has never been lower than it is right now. In 1935,
there were roughly 450 deaths for every 100,000 American kids
between 1 and 4 years old. Today, that number rarely reaches 30.
Mortality rates are down by almost half since 1990, and for a kid
between 5 and 14 years old, the chances of premature death are
around 1 in 10,000, or 0.01 percent, according to reporting by

Christopher Ingraham in the *Washington Post*, who concludes, "Kids are dying less. They're being killed less. They're getting hit by cars less. And they're going missing less frequently, too. The likelihood of any of these scenarios is both historically low and infinitesimally small. . . . Bottom line: If it was safe enough for you to play unsupervised outside when you were a kid, it's even safer for your own children to do so today."

While some of these reductions in risk to kids are probably due to increased parental vigilance, there's little doubt that while parents are in a frenzy trying to protect their children, they may be distracted from far more potent threats such as lack of exercise, unhealthy diet, chronic stress, too little face-to-face interaction with friends, and lack of free time and access to nature—all of which are taking a horrible toll on children.

We keep them indoors because we think it's safer, but dangers lurk in the house. Inactivity leads to obesity and diabetes, but other, less obvious conditions can be traced to overly sheltered childhood as well. In the past five decades, for example, the number of young adults in the United States and Europe with myopia has doubled, from about a quarter to half. Some researchers are predicting that by 2020, one-third of the world's population could be diagnosed with the condition. Why? Partly because kids aren't getting enough sunlight in their retinas and are focusing almost entirely on things that are close by, both because they're spending so much of their time indoors.

Our panicky need to protect kids from imaginary dangers may owe something to the notion that the welfare of children is solely the responsibility of the parent, and not of the community, the extended family, or the child herself. Liedloff's description of the Yequana's approach to parenting along the Orinoco River in the Venezuelan Amazon echoes that of most anthropologists who've spent time among foragers: "The notion of ownership of another person is absent among the Yequana. The idea that this is 'my child' or 'your child' does not exist. Deciding what another person should do, no matter what his age, is outside the Yequana vocabulary of behaviors."

The growing anxiety around parenting in the United States may also be tied to economic inequality. Fabrizio Zilibotti and Matthias Doepke are economists whose research on parenting is explained in their book *Love, Money, and Parenting: How Economics Explains the Way We Raise Our Kids*. They found that, compared to a generation or two ago, the amount of time parents spend supervising their kids has risen dramatically—especially in countries where economic inequality has also been increasing. As Zilibotti explains, "In a society that is very unequal—where there are lots of opportunities if one does well and very negative outcomes if one is less successful—parents will be more worried that their children won't become high achievers in school. But if you go to a country where there is less inequality, parents may be less worried about that, not because they care less about their children, but because the negative outcomes aren't as bad." Other considerations, such as the children's happiness and individuality, can be sacrificed to the frenzy to succeed.

Given the unrealistic expectations placed upon both parents and children, and the American tendency to see misalignments between our evolved nature and our current society as pathologies that can be addressed with pharmaceuticals, it shouldn't be surprising that we're drugging kids into lethargic submission. By high school, nearly one in five boys in the United States will have been diagnosed with ADHD—a "disease" that strikingly resembles normal juvenile primate behavior: a need for plentiful physical activity, skepticism of authority figures, an insatiable hunger to play. In 1997, the Centers for Disease Control estimated that around 3 percent of American schoolchildren had been diagnosed with ADHD. By 2013, the percentage had exploded to 11 percent, and an astonishing 15.1 percent for boys. And of those who've received the diagnosis, two-thirds are being given prescription drugs. According to the manufacturers' warnings, side effects can include sudden death in children with undiagnosed heart problems, bipolar conditions, aggressive behavior or hostility, psychotic symptoms (such as hearing voices and believing things that are not true), manic symptoms, facial tics, sleep disorders, paranoia, and suicidal feelings. And yet, sales of ADHD drugs increased by 89 percent between 2008 and 2016, rising from $5.5 billion to an estimated $12 billion to $14 billion. We seem to have decided that it's too expensive or inconvenient to modify the environments our children learn in, so we're modifying their brain chemistry instead.

The suspicion that a lot of kids are being drugged just for being kids is supported by a study published in the *Canadian Medical Association Journal* showing that boys born in December (thus, typically the youngest boys in their class) "were 30 percent more likely to receive a diagnosis of ADHD than boys born in January," and these boys were 40 percent more likely to be given a prescription for meds. Their "sickness" appears to boil down to having been born in December instead of January.

Developmental psychologist Peter Gray has written extensively about how foragers consider children deserving of respect: "Hunter-gatherers' treatment of children is very much in line with their treatment of adults. They do not use power-assertive methods to control behavior; they believe that each person's needs are equally important; and they believe that each person, regardless of age, knows best what his or her needs are." Gray then links this individual autonomy to the ecological and economic context of the hunter-gatherer social world, noting that "children are not dependent on any specific other individuals, but upon the band as a whole, and this greatly reduces the opportunity for any specific individuals, including their parents, to dominate them . . . children are free to move into other huts—most commonly the huts of their grandparents or uncles and aunts—if they feel put upon by their parents."

Training the desire for play out of children is a bit like teaching birds not to sing; it can probably be done, but why? Kids play because it teaches them how to live together. Gray has concluded that play is not only fundamental to the cognitive and physical development of children but was also "a foundation for hunter-gatherer social existence." Gray sees "play and humor . . . at the core of hunter-gatherer social structures and mores." They function to promote an "egalitarian attitude, extensive sharing, and relative peacefulness for which hunter-gatherers are justly famous and upon which they depend for survival." Play, in Gray's view, amounted to serious business in foraging communities, in that it "provided a foundation for . . . modes of governance, religious beliefs and practices, approaches to productive work, and means of education."

But increasingly, even the smallest children's lives are being oriented away from play and toward work. Daphna Bassok, a researcher specializing in educational policy, found that in 1998, 30 percent of American teachers believed that children should learn to read while in kindergarten. By 2010, that figure had almost tripled, to 80 percent. The absence of time to just hang out and play together is having serious consequences in how kids develop. "They can do math in first grade, but they are not attuned to subtle social cues," says Dr. Ellen Littman, a clinical psychologist and coauthor of *Understanding Girls with ADHD*. "They are not developing the normal skills that come from interacting with play, including how to manage their emotions." Peter Gray agrees. "Where do children learn to control their own lives? When adults aren't around to do it for you," he said. "If you don't have the opportunity to experience life on your own, to deal with the stressors of life, to learn in this context of play where you are free to fail, the world is a scary place."

Chapter 8

Turbulent Teens

Women seem wicked when you're unwanted . . .

—The Doors, "People Are Strange"

I was an angry teen. Like many in the civilized world, I felt something pulling me toward a long slog of regimentation, meaningless work, and ever-increasing isolation. I suspect that the anger of many teens is fueled by the same kind of dismay at the dawning realities of adulthood.

Another reason I was angry, to be honest, is that my burgeoning sexual awareness had become a source of frustration, shame, and confusion. As the hormonal surge swept through me, the possibility of exploring this new world with a girl or woman became increasingly urgent and unlikely. There was something deeply unjust about needing something so badly (sex? love? intimacy? touch?) while the practical conditions of life made the chances of finding it just about zero. We laugh at the sexual frustrations of testosterone-addled, pimply-faced, braces-wearing geeks in movies—because, well, they're in a laughable situation. But their suffering is real, and the intense frustration and humiliation experienced by young people who feel they're being denied something they need at the core of their being generates a dangerous pressure. In young men, this pressure all too often explodes outward in misogyny, rage, and violence, while in young women it tends to implode, manifesting as depression, self-harm, and eating disorders.

Griffin Hansbury, a female-to-male transsexual, offered some rare insight into these frustrations when he spoke with Alex Blumberg on *This American Life*. Hansbury explained how it felt to suddenly be swept up in a rising tide of testosterone:

> The most overwhelming feeling is the incredible increase in libido and change in the way that I perceived women and the way I thought about sex. Before testosterone . . . I would see a woman on the subway, and I would think, she's attractive. I'd like to meet her. What's that book she's reading? I could talk to her. This is what I would say. There would be a narrative. There would be this stream of language. It would be very verbal. [But] after testosterone, there was no narrative. There was no language whatsoever. It was just . . . aggressive, pornographic images, just one after another. It was like being in a pornographic movie house in my mind. And I couldn't turn it off. I could not turn it off.

Hansbury said he "felt like a monster a lot of the time," but he gained a great deal of compassion for men and boys. "It made me understand adolescent boys a lot," he said, before recounting an experience I think every straight adolescent boy can relate to:

> I remember walking up Fifth Avenue. There was a woman walking in front of me. And she was wearing this little skirt and this little top. And I was looking at her ass. And I kept saying to myself, don't look at it, don't look at it. And I kept looking at it. And I walked past her. And this voice in my head kept saying, turn around to look at her breasts. Turn around, turn

around, turn around. And my feminist, female background kept saying, don't you dare, you pig. Don't turn around. And I fought myself for a whole block, and then I turned around and checked her out.

There are many legitimate opinions on how a society should manage these impulses, but few things inspire murderous mayhem in male human beings more reliably than sexual repression. When the free expression of sexuality is thwarted, the human psyche tends to grow twisted into grotesque, enraged perversions of desire. Recall James Prescott's research, mentioned previously, in which two measures of affectional bonding (mother-infant contact and the free exercise of youthful sexual expression) predicted with 100 percent accuracy whether forty-nine tribal cultures would be violent or peaceful. Unfortunately, the distorted rage resulting from the repression of youth sexual exploration rarely takes the form of rebellion against the people and institutions behind the prohibitions. (If it did, perhaps we'd be reading of abused priests rather than priests as abusers.) Instead, the rage is generally directed at the self (as shame), or at the girls and women who are misperceived as being the cause of the frustration. Entire cultures seem determined to find women to blame for whatever befalls humanity. In 2010, the BBC quoted Kazem Sedighi, an Iranian cleric, saying, "Many women who do not dress modestly lead young men astray and spread adultery in society which increases earthquakes." Iranian clerics not being known for their impish sense of humor, I think we can assume this was said with a straight face.

Christianity is a religion centered upon a figure who was supposedly conceived asexually by a virgin mother. Sexual hang-ups, anyone? Mark Twain noted the striking antieroticism of Christianity as expressed in its bizarrely sexless heaven: "[Man] has imagined a heaven, and has left entirely out of it the supremest of all his delights, the one ecstasy that stands first and foremost in the heart of every individual of his race . . . sexual intercourse! It is as if a lost and perishing person in a roasting desert should be told by a rescuer he might choose and have all longed-for things but one, and he should elect to leave out water!"

There's little question that the centuries-long campaign of child rape enabled by institutional cover-up is a direct result of the Church's inhumane denial of human sexuality. Gay, conservative, Catholic author Andrew Sullivan has written movingly about the corrosion of the human spirit required to deny our deepest sexual nature:

> Accepting God's unconditional love for me was the hardest part of keeping hold of my Christian faith. My childhood and adolescence were difficult to the point of agony, an agony my own church told me was my just desert. But I saw in my own life and those of countless others that the suppression of these core emotions and the denial of their resolution in love always *always* leads to personal distortion and compulsion and loss of perspective. Forcing gay people into molds they do not fit helps no one. It robs them of dignity and self-worth and the capacity for healthy relationships. It wrecks family, twists Christianity, violates humanity. It must end.

Of course, it's not just a question of repressing homosexuality, but of repressing *all* sexuality. *Homo sapiens* is a deeply, essentially sexual species. For millennia, institutions have roped us into a conspiracy of shame in which we pretend our sexual energies are easily ignored and tamed—while most of us know that's not true in our personal experience, so we must be sick. Nearly all of us harbor the shameful secret we dare not share.

Religions aren't the only institutions to leverage this abuse of spirit and body. Medical doctors have long participated in some of these crimes against our humanity. In 1850, the *New Orleans Medical and Surgical Journal* declared masturbation public enemy number one, warning: "Neither plague, nor war, nor smallpox, nor a crowd of similar evils, have resulted more disastrously for humanity than the habit of masturbation: it is the destroying element of civilized society." "Scientific" declarations like these encouraged Dr. John Harvey Kellogg (brother of the Corn Flakes Kellogg) in his campaign to eradicate masturbation from the United States. Though one of the leading sex educators of his day, Kellogg claimed never to have had intercourse with his wife in over four decades of marriage. (One wonders: In what realms other than sexuality and drug policy would the *total absence* of relevant personal experience be seen as a glowing qualification for giving advice?)

In his bestseller *Plain Facts for Old and Young*, Kellogg outlined the optimal ways to dissuade children from masturbating. In a chapter called "Treatment for Self-Abuse and Its Effects," he recommended circumcision as "a remedy which is almost always successful in small boys," specifying that "the operation should be performed by a surgeon *without administering an anesthetic*, as the brief pain attending the operation will have a salutary effect upon the mind, especially if it be connected with the idea of punishment" (emphasis added). Further, Kellogg recommended "the application of one or more silver sutures in such a way as to prevent erection. The prepuce, or foreskin, is drawn forward over the glans, and the needle to which the wire is attached is passed through from one side to the other. After drawing the wire through, the ends are twisted together and cut off close. It is now impossible for an erection to occur." Not to worry. Kellogg assured parents that sewing their son's penis into its foreskin "acts as a most powerful means of overcoming the disposition to resort to the practice [of masturbation]." This was nothing less than institutionalized, culturally sanctioned child abuse. The fact that healthy young men typically have several erections every night while dreaming suggests the incalculable trauma caused by this twisted man and his pathological advice.

Girls were not spared Kellogg's toxic tortures. In the same book he advises parents to pour carbolic acid on the clitoris of little girls who are found to be touching themselves inappropriately. All this suffering was called for because "science" had proven that masturbation caused impotence, testicular atrophy, uterine disease, sterility, heart disease, epilepsy, blindness, deafness, idiocy, and insanity. It took American physicians nearly a century to openly question these absurd notions, and, even now, circumcision—which is rarely a medical necessity—remains prevalent in the United States. As sexologist John Money explained, "Neonatal circumcision crept into American delivery rooms in the 1870s and 1880s, not for religious reasons and not for reasons of health or hygiene, as is commonly supposed, but because of the claim that, later in life, it would prevent irritation that would cause the boy to become a masturbator."

While data on teenage masturbation among foragers is scant, anthropologists, explorers, and flustered missionaries have all reported casual sexual exploration among young people as being tolerated, when not encouraged, by adults. Among the !Kung San, for example, Konner reported that "adults . . . considered sexual experimentation in childhood and adolescence to be inevitable and normal." Indeed, among the !Kung San "sexual activity [is] considered essential for mental health, and [they] sometimes referred to mentally ill people . . . as deranged because of sexual deprivation." One of America's bravest anthropologists, Margaret Mead, caused an uproar when she reported that children and adolescents on the South Pacific islands she'd studied experimented freely with sex. For them, good sexual compatibility was the prerequisite for intimacy: "Personal affection may or may not result from acts of sexual intimacy," Mead reported, "but the latter are requisite to the former—exactly the reverse of the ideals of western society."

In a chilling and prescient essay called "The Weaponized Loser," Stephen T. Asma, a philosopher, explains that the "fear and loathing of emancipated female sexuality" is effectively converted into a recruitment tool for young jihadis. But Asma is not interested only in attacks originating in repressive Muslim cultures. He begins his essay, published in 2016, by noting that, "Of the past 129 mass shootings in the United States, all but three have been men. The shooter is socially alienated, and he can't get laid." In light of those stark facts, mass shootings in the United States appear to be analogous to terrorist operations in a homegrown jihad fueled by the sexual shame permeating American culture. Think I'm overstating it? Asma argues that American culture essentially teases these young men by intensifying their pre-existing desires with constant references to and depictions of sex: "These are existential issues because they resonate—rightly or wrongly—at the core of how many men see themselves." This process, he argues, increases their resentment, "the psychological fuel that gets the fire of violence going, whatever the ideological justification."

I'm not arguing that sexual frustration alone is responsible for jihad or American mass shootings, but there's a reason martyrs are promised seventy-two virgins and why Christopher Harper-Mercer left a note lamenting that "I am going to die friendless, girlfriendless, and a virgin," before shooting a professor and eight fellow students in Oregon. Similarly, before killing six people, and wounding thirteen more, twenty-two-year-old Elliot Roger uploaded a YouTube video where he said that he "felt a new sense of power" when he picked up the handgun: " 'Who's the alpha male now, bitches?' I thought to myself, regarding all the girls who've looked down on me in the past."

As chance would have it, I was driving through Roseburg, Oregon, with a friend from Holland when Christopher Harper-Mercer died a virgin at nearby Umpqua Community College. We were listening to the Doors when dozens of police cars screamed past us, seemingly coming from every direction: "People are strange when you're a stranger. Faces look ugly when you're alone. Women seem wicked when you're unwanted . . ." We switched to the radio to find out what was happening. When we heard about the shootings, my friend Martijn said, "Welcome to America." Of course, terrible things happen in Holland, but not *this* sort of terrible thing. I asked Martijn what was different about the way the Dutch handle teenage sexuality that enables them to avoid this kind of blind rage. He suggested I look into Dutch sex education programs, which, he assured me, "treat kids as sexual beings, and respectfully."

He's right. Sociologists Jane Lewis and Trudie Knijn studied these programs and found that they were far more likely than such programs in other countries (specifically, England and Wales) to cover such potentially controversial topics as female sexual pleasure, homosexuality, and masturbation. The Dutch programs emulate many of the ways foragers approach childhood development, with their emphasis on mutual respect and individual autonomy in negotiating adolescent sexual relationships. These programs have been astoundingly successful, however you assess them. Sociologist Amy Schalet conducted a survey of Dutch youth between twelve and twenty-five and found that "the majority described their first sexual experiences—broadly defined—as well-timed, within their control, and fun. About first intercourse, 86 percent of women and 93 percent of men said, 'We both were equally eager to have it.'" The rate of births to Dutch teen mothers is consistently among the lowest in the world, as is the abortion rate for Dutch girls. In 2007, the rate of births to American girls between fifteen and nineteen was eight times higher than that of Dutch girls of the same age group. Since the 1980s, American "abstinence only" programs have received lavish federal funding, despite their obvious failure to address the reality of teenage sexuality, resulting in the highest teen pregnancy rates in the industrialized world.

This inanity is fueled both by the so-called American Taliban— sex-phobic Christian fundamentalists who have far more political leverage than justify their numbers—and by complicit American parents who refuse to face the reality of their children's sexuality. Meanwhile, a study of Dutch parents by Janita Ravesloot found that, in most families, youth sexuality was accepted as simply being part of a young person's life, "from first kiss to first coitus." A 2003 study found that two-thirds of Dutch kids from fifteen to seventeen were allowed to have their girlfriend or boyfriend spend the night with them, in their bedrooms at home.

In her research, Schalet found that American parents consistently view their kids' emerging sexuality through a Hobbesian lens, emphasizing the "dangerous and conflicted elements," the "raging hormones," and supposedly innately antagonistic relation between the sexes, with girls pursuing love as boys are fixated only on sex. "Viewing sex as part of a larger tug of war between separation and control," writes Schalet, "the response to the question of the sleepover, even among otherwise socially liberal [American] parents is, 'Not under my roof!'"

At this point, I imagine many American readers will be

exasperated with me, a nonparent, suggesting that some of their struggles with teenagers are unnecessary. "You don't understand. Teenagers are lunatics! They *need* to rebel." I hear you. But the data suggest teens are rebelling against something in particular—namely, a culture they find oppressive and unfair—even if they rarely articulate their rage in those terms. When the character played by Marlon Brando in the 1953 classic film *The Wild One* is asked, "Hey, Johnny, what are you rebelling against?" he responds, "Whadda you got?" In other words, "All of it."

But this undifferentiated rage isn't a universal characteristic of teenagers. Cross-cultural evidence strongly suggests that the difficult period we call adolescence is in fact a recent cultural artifact. When anthropologists Alice Schlegel and Herbert Barry III reviewed research on teens in the 186 preindustrial societies for which data were available, they found that more than half these cultures didn't even have a word for "adolescence." Teens in these cultures showed almost no signs of psychopathology, and antisocial behavior among young males was completely absent in more than half the cultures and extremely mild in the rest. An associated study found that problems tied to teen rage didn't begin to appear until shortly after the introduction of Western influences, especially schooling and media. In contrast, in 2015, about three million American teens between twelve and seventeen years old had at least one major depressive episode, according to the Department of Health and Human Services—and experts suspect these numbers are low, due to many unreported cases. According to the National Institute of Mental Health, roughly 30 percent of girls and one in five boys have suffered from an anxiety disorder. "If you wanted to create an environment to churn out really angsty people, we've done it," said Janis Whitlock, director of the Cornell Research Program on Self-Injury and Recovery.

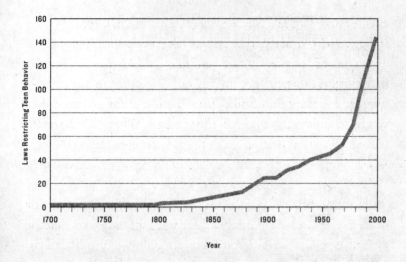

In his 2007 essay in *Scientific American*, "The Myth of the Teen Brain," psychologist Robert Epstein argues that "the turmoil we see among teens in the U.S. is the result of . . . 'artificial extension of childhood' past puberty. Over the past century," he points out, "we have increasingly infantilized our young, treating older and older people as children while also isolating them from adults." After looking at laws and regulations restricting teenagers' behavior, Epstein found that "teens in the U.S. are subject to more than 10 times as many restrictions as are mainstream adults, twice as many restrictions as active-duty U.S. Marines, and even twice as many restrictions as incarcerated felons." Additional research conducted by Epstein and Diane Dumas demonstrated a significant link between "the extent to which teens are infantilized and the extent to which they display signs of psychopathology. . . . There is no question that teen turbulence is *not* inevitable," Epstein concludes. "It is a creation of modern culture, pure and simple—and so, it would appear, is the brain of the troubled teen."

If modern culture is behind the brain of the troubled teen, what's it doing to the brain of the troubled adult?

Chapter 9

Anxious Adults

– Good Work, If You Can Get It –

I tell you, we are here on Earth to fart around, and don't let anybody tell you different.

—Kurt Vonnegut

If any man would not work, neither should he eat.

—2 Thessalonians 3:10

When documentary filmmaker Jonnie Hughes was living with the "Insect Tribe" in a remote part of Papua New Guinea, a few of the tribesmen who had been hosting him asked Jonnie if they could visit him back in the United Kingdom. A few months later, when Jonnie pitched the idea of flying a few foragers to London, his bosses saw the documentary value and agreed to fund their trip. But Hughes was worried the visit might "pollute their culture with modern ideas, or perhaps make them terminally envious of a world beyond their reach." After all, these were people who were living in very primitive conditions, with no refrigeration, modern medicine, television, or other marvels of modernity. By the end of the visit, however, Hughes saw things very differently:

With every whispered observation, they left us powerless to explain the madness of our own social norms, and when they boarded the plane back to PNG, we were the ones racked with envy—envious of their joyously interdependent community, their clear understanding of what mattered in life, their rock-solid roles, simple pleasures and ample leisure time, their lack of mortgages and debts, their indisputable "goodness." Our world appeared an obscene and dysfunctional manifestation of human existence in comparison.

If Hughes sounds a bit like one of those silly romantics we're always being warned about, just do the numbers. Hughes says the tribesmen "were fascinated about our work/life balance, because over there, in a week, they'll spend maybe twenty hours in total collecting food, going hunting, *etc.*—just doing the things they need to do. The rest of their time they spend with their family, social lives . . . leisure time."

No wonder they were confused that Mark, the father in the family they were staying with, left early every morning and didn't return until evening. "Why are you doing this?" Hughes recalls them asking. "Why are you going out every day, not seeing the people that you really care about? It doesn't make any sense at all!" Mark explained that he had to work to pay for the house they were living in. "How long will you be doing this, to pay for your house?" they asked. When Mark told them about his twenty-five-year mortgage, they looked at him in astonished pity, explaining that when one of them needed a house, they got together with the other men of the village and built a house in a couple of weeks.

At the end of their visit, the Insect People took just one innovation back to Papua New Guinea: the notion of putting feathers on arrows to stabilize their flight. Apparently, that was the only thing that impressed them very much about the modern world.

Anthropologists long ago established that almost without exception, hunter-gatherers rarely "work" more than three or four hours per day, and these activities are "integrated with rituals, socialization, and artistic expression to a degree unknown to most people in Western societies," as John Gowdy explains. Gowdy, an economist, edited a collection of essays by anthropologists and economists called *Limited Wants, Unlimited Means*, showing many ways in which the behavior of foragers is consistently *the opposite* of what modern economic theory assumes to be universal among all human beings.

"The idea that work is drudgery whose only purpose is to enable people to live their 'real' lives is not present in hunter-gatherer societies. The work-leisure trade-off discussed in economic textbooks is apparently absent there," according to Gowdy. There is no word for "work" in the Yequana language, says Liedloff, which makes sense in that the concept was foreign to them. "There were words for each activity that might have been included, but no generic term [for work]." Peter Gray agrees that "in general, hunter-gatherers do not have a concept of toil." Gray sees a continuity between children's play and adult activities that make the notion of "work" incomprehensible. He notes that kids play at hunting, gathering, toolmaking, and all the other activities of adults, just as puppies play at the activities of wolves. As they grow older, Gray writes, "play becomes work, but it does not cease being play. It may even become more fun than before, because the productive quality helps the whole band and is valued by all."

Gray has identified four main reasons to conclude that the daily activities necessary to a forager's survival are more accurately seen as play than as work:

1. It is varied and requires skill and intelligence, leading to the satisfaction of being good at something that requires

our full attention. "The [foragers'] abilities include physical skills, honed by years of practice, as well as the capacity to remember, use, add to, and modify an enormous store of culturally shared verbal knowledge."

2. There is not too much of it. Anthropologists in various parts of the world have determined that foragers rarely "work" more than a few hours per day.

3. It is done in a group of friends. In most ecological contexts, hunting parties consist of at least a few men, and women almost always hunt and gather in groups. Anthropologist Alf Wannenburgh described gathering expeditions with the !Kung San as "jolly events," often having "something of the atmosphere of a picnic outing with children."

4. It is, for any given person at any given time, optional. This is the most important aspect determining the "playfulness" of the hunter-gatherer existence, as it is the sense of choice that ultimately determines whether something is work or play. Crucially, foragers have found a way to get things done while maximizing every participant's sense of being free to join or not.

Raised in a world organized around the concept of scarcity, we find it hard to imagine that our ancestors (supposedly locked in eternal struggle for mere survival) found a way to unlink gain from pain, but anthropologists have confirmed that there is little or no connection between who produces and who receives the economic output in many foraging societies studied. Disdain for unproductive people makes sense in societies that consider the activities necessary for food and shelter to be arduous and disagreeable. After all, if work is hard, why should I do more than you? But if those activities are the sorts of things we enjoy doing in our free time (hunting, walking, fishing, repairing a hut, playing with children), that logic falls apart. If hunting, for example, is fun, why should those who choose to do it the most be held in higher regard than anyone else?

Foragers are rarely willing to join the ranks of the employed until they are forced into it. Even then, they are notoriously unindustrious. Speaking of the Yamana people of Tierra del Fuego, German ethnologist Martin Gusinde lamented that:

> The Yamana are not capable of continuous, daily hard labor, much to the chagrin of European farmers and employers for whom they often work. Their work is more a matter of fits and starts, and in these occasional efforts they can develop considerable energy for a certain time. After that, however, they show a desire for an incalculably long rest period during which they lie about doing nothing, without showing great fatigue. . . . The Indian cannot help it. It is his natural disposition.

Is this "natural disposition" best seen as laziness or just a healthy disdain for meaningless labor?

On the other side of the world, some of the first Europeans to live in Australia felt compassion for the "miserable Aborigines" who were "reduced by famine to the miserable necessity of subsisting on certain kinds of food, which they have found near their huts," including insects, rodents, and larvae. Although the Europeans noted the native people's apparent health, happiness, and great appetite for lying about in hammocks, it seems never to have occurred to them that the native people were eating nutritious and plentiful food that was readily available without work.

In our world, work is ubiquitous. The server asks me if I'm still "working on" my salad. We don't exercise, we "work out." Learning about ourselves in therapy is referred to as "doing the work." "What are you working on?" has become a way of asking who you are. Yet few of us are lucky enough to work in ways that actually align with who we are. We roll in, cup of stimulant in hand, shuffle papers, try to look busy and stay awake, fight the hopelessness, then go home and drink too much. Show up, punch in, tune out. In his semiautobiographical novel, *The House of the Dead*, Fyodor Dostoyevsky wrote, "If one wanted to crush and destroy a man entirely, to mete out to him the most terrible punishment . . . all one would have to do would be to make him do work that was completely and utterly devoid of usefulness and meaning." But from factory floor to corporate boardroom, useless, meaningless work is standard-issue in our world. And you're expected to be grateful to have it!

No wonder a recent poll conducted by the Harvard Institute of Politics found that fewer than one in five Americans between eighteen and twenty-nine considered themselves "capitalists." Only 42 percent of them said they "supported capitalism." As *Time* magazine reported, "This represents more than just millennials not minding the label 'socialist' or disaffected middle-aged Americans tiring of an anemic recovery. This is a majority of citizens being uncomfortable with the country's economic foundation—a system that over hundreds of years turned a fledgling society of farmers and prospectors into the most prosperous nation in human history."

But how can we say a nation is "the most prosperous" in history when its infrastructure is collapsing, its mentally ill are condemned to prisons, millions are denied even basic medical care, one in five children go to bed hungry each night, and so on? What does "prosperity" mean in a country where 47 million people are below the official poverty line, and millions more hover just above it? It's unconscionable to follow the common but absurd pattern of averaging the astronomical wealth of a few families into comforting, meaningless statistics and calling the United States "prosperous." Preposterous, perhaps, but certainly not prosperous.

In any case, prosperity isn't the key to life satisfaction. Italian economist Paolo Verme found the variable "freedom and control" to be the most significant predictor of self-reported quality of life, by far. The kind of freedom that leads most directly to happiness, in other words, is the freedom *not* to get up to the ringing of an alarm five days a week, *not* to be obligated to shave and put on a tie (or bra) if you don't feel like it, *not* to pretend to respect someone you don't just because he's your "boss" just so you'll have enough money to keep the bill collectors at bay for another month.

In 1932, the philosopher Bertrand Russell published a charming, brilliant essay called "In Praise of Idleness," in which he noted that "the morality of work is the morality of slaves, and the modern world has no need of slavery." If an Oxford University study is correct in projecting that 47 percent of all jobs in the United States will be lost to automation by 2030, we'll soon hardly need work, much less slavery. Almost a century ago, Russell recognized that most of the hours human beings spend working are a total waste of time, noting that "only a foolish asceticism, usually vicarious, makes us continue to insist on work in excessive quantities now that the need no longer exists." He ties the problem to the facts that "the idea that the poor should have leisure has always been shocking to the rich" and that the industrial labor mobilized for World War I was never demobilized. Of course, a decade after his essay was published, another mobilization was under way, on a much larger scale, ultimately congealing into what President Eisenhower called "the military-industrial complex." What's most striking about Russell's essay is its last paragraph, in which he envisions a future for humankind that sounds nearly identical to our prehistoric past:

> Above all, there will be happiness and joy of life, instead of frayed nerves, weariness, and dyspepsia. The work exacted will

be enough to make leisure delightful, but not enough to produce exhaustion. . . . Ordinary men and women, having the opportunity of a happy life, will become more kindly and less persecuting and less inclined to view others with suspicion. The taste for war will die out, partly for this reason, and partly because it will involve long and severe work for all. Good nature is, of all moral qualities, the one that the world needs most, and good nature is the result of ease and security, not of a life of arduous struggle. . . . Hitherto we have continued to be as energetic as we were before there were machines; in this we have been foolish, but there is no reason to go on being foolish forever.

If work is unnecessary, why do we continue to behave as if the key to a good life is to spend most of it doing something we'd rather not?

– The Price of Money –

Money often costs too much.

—Ralph Waldo Emerson

He who dies with the most toys wins.

—Malcolm Forbes

The more we understand what human life was like before agriculture, the more civilization looks like a pyramid scheme. Disparities of wealth and power were among the first things to emerge when people settled into villages and towns. *Someone* had to make decisions about who got how much of what, and when. *Someone* had to organize the sowing and the reaping, the protection and trading of land and livestock. Once wealth emerged, so did an elite class that was naturally tempted to benefit further from their privileged position.

When similar situations arise in foraging societies—when a large animal is killed, for example—formal codes of behavior kick in to prevent inequities in the distribution of the windfall. Among bands of a few dozen foragers who all know each other intimately, cheating is quickly detected and discouraged—initially with lighthearted humor, but with the serious threat of more severe repercussions if a light ribbing proves ineffective.

Once human communities grew beyond the point where every individual had a direct relationship with everyone else, something fascinating and terrible happened: Other people became abstractions. Perhaps Joseph Stalin was thinking along these lines when he said, "One death is a tragedy; a million is a statistic." When the number of human beings rises to the point where it's no longer possible to picture the faces of those who are affected by our decisions, innate human compassion is often overwhelmed by other concerns. Politicians who would unthinkingly jump into a river to save a drowning child have no qualms about approving policies that leave millions of impoverished kids floundering without basic health care or school lunches. Humans seem to be two different creatures when we compare how we function in small-scale versus large-scale societies. Grasshoppers and locusts.

Wealth disparities unimaginable to foragers are common in the modern world. In the United States, wealth distribution hasn't been this out of whack since the so-called Roaring Twenties. In 2012, according to research compiled by French economist Thomas Piketty and his colleagues, the top 1 percent of households in the United States took 22.5 percent of total income, the highest proportion since 1928. In the 1950s, an American CEO could expect to be paid about twenty times more than a typical worker at his firm. Today, the ratio is more than ten times that—over two hundred to one. And some CEOs make that kind of ratio look downright Marxist. In 2011, Apple's Tim Cook was paid $378 million in salary, stock, and other benefits—6,258 times the wage of the average employee at Apple. *The richest eighty-five people in the world control more wealth than the poorest half of the planet's population.* Let that sink in for a moment. Eighty-five human beings who fart in bed just like you and me control more wealth than *3.5 billion* other people—many of whom live in desperate poverty. Piketty, who is "arguably the world's leading expert on income and wealth inequality," according to Nobel laureate Paul Krugman, has concluded that income inequality in the United States today is "probably higher than in any other society at any time in the past, anywhere in the world." Such disparities of wealth are not just inhumane, they are inhuman, offending our innate predisposition for fairness.

When three Tupinambá natives were taken to France from Brazil in the sixteenth century, the essayist Montaigne was present at their visit with King Charles IX. When the natives were asked what they found most peculiar about the European way of life, Montaigne recounts, "they had observed, that there were among us men full and crammed with all manner of commodities, while, in the meantime, [others] were begging at their doors, lean, and half-starved with hunger and poverty; and they thought it strange that these necessitous [people] were able to suffer so great an inequality and injustice, and that they did not take the others by the throats, or set fire to their houses."

Of course, sometimes poorer people *do* rise up and set fire to the houses of the rich, but things soon settle into the same pattern of an elite few profiting from the labor of the unorganized masses, yet again. Meet the new boss, same as the old boss. Given the recurrent pattern, it's not surprising that many have concluded that this state of affairs is simply the result of human nature—or of nature itself. Many of the great robber barons of the twentieth century were fond of twisting Darwin's theories to imply that their wealth was simply the logical result of their innately superior "fitness," and was therefore as natural and inevitable as any other form of predation upon the weak by the strong. In his essay "Gospel of Wealth," for example, Andrew Carnegie argued that while this "natural law" leads to great suffering among the poor, "it ensures the survival of the fittest in every department. . . . We accept and welcome, therefore, as conditions to which we must accommodate ourselves, great inequality of environment, the concentration of business, industrial and commercial, in the hands of a few, and the law of competition between these, as being not only beneficial, but essential for the future progress of the race."

But while Darwin believed economic inequality to be a necessary first step in the development of civilization, he knew that material inequality wasn't present in many of the societies he'd visited in his travels, and that such inequality must, therefore, be something more complicated than a straightforward expression of human nature. Darwin's observations have been confirmed by contemporary researchers. Gowdy concludes that "all the assumptions economists make about economic man are absent in [foraging] societies. People in immediate-return societies are not acquisitive, self-centered cost-benefit calculators. In these societies, it can be most clearly seen that economic man as a universal human type is a fiction." The more we learn about foragers, the clearer it becomes that their lives are more approximate expressions of human nature than ours are, in that modern market capitalism requires an array of subversions of our natural behavior. "The view of human nature embedded in Western economic theory is an anomaly in human history," Gowdy concludes. "The hunter-gatherer represents 'uneconomic man.'"

At an anthropology conference in 1966 called "Man the Hunter," Marshall Sahlins presented research that posed the first substantive modern-day challenge to the Hobbesian paradigm of prehistoric life. In a symposium called "The Original Affluent Society," Sahlins introduced many of the ideas I've been arguing in these pages. A few years later, he articulated his thesis in more detail in a book called *Stone Age Economics*, in which he wrote, "The world's most primitive people have few possessions, *but they are not poor*. Poverty is not a certain small amount of goods, nor is it a relation between means and ends; above all it is a relation between people. Poverty," Sahlins declared, "is a social status. As such it is the invention of civilization." Israeli anthropologist Nurit Bird-David went a step further, arguing that foragers aren't merely *not poor*; their behavior suggests they believe themselves to be rich: "Just as Westerners' behaviour is understandable in relation to their assumption of shortage, so hunter-gatherers' behaviour is understandable in relation to their assumption of affluence." Noble savages, indeed.

– How to Lose by Winning –

People who say the system works work for the system.

—Russell Brand

If poverty is a relative concept, so is wealth. Counterintuitively, in the civilizational game, the biggest winners are often total losers. I'm not arguing that the criminally skewed wealth distribution in the modern world should be excused, forgiven, or ignored. And

I'm certainly not forgetting the brutal fact that while billions of people scavenge for their next meal or some clean water, a few live in hilltop mansions pouring last night's flat champagne down the drain. But the accelerating processes by which our species is transforming this planet from the wonder of wonders to "an immense pile of filth," in Pope Francis's words, benefit the super-wealthy only in limited ways, and only for a while. It's true that they'll never have to worry about starving, finding a job, or raising a family in the back seat of their Lamborghini, but they can't buy their way out of the storms we all face. Rising seas don't distinguish mansions from shacks. The wealthy and their children breathe the same fouled air, bathe in the same toxic water, and eat food steeped in the same poisons and cruelty. A stressed-out millionaire may get the best chemotherapy money can buy, but he's still going to get the cancer. The rich are ultimately subject to the same rules of nature as everyone else.

Money is like food, rain, wives, husbands, kids, cats, sex, TV stations, and decorative pillows in that more than enough is too much. But because we're so indoctrinated to believe that money is the golden exception to the rule of diminishing returns, it's very difficult to know when to stop striving for more, to take the money and run.

Years ago a man sitting next to me on a train in India explained how his grandfather had hunted monkeys in the hills north of Calcutta. He made a small wooden box with a round hole in the side. Before attaching the top, he placed a mango in the box, then strapped it to a tree, where a passing monkey would smell the rotting mango and reach into the box through the hole. But mango pits are too large to pull out through the hole. So the monkey faced a dilemma: let go of the mango and be on his way, or sit there, holding the uneaten fruit, until the hunter came along to shoot him. The traps, the man said, were very effective.

Who among us has the good sense to drop the mango and walk away? I know, you *think* you'd buy a cozy cottage and chillax if you had a million dollars in the bank, but would you really? Once you had that million, you'd no longer be the person you are today. You'd have a different group of friends—many of whom had a lot more than a million dollars tucked away. Your "normal" would have shifted to something a lot more expensive to maintain. The cues in your environment telling you what "normal" means would be sending new, more expensive signals.

I like red wine a lot—maybe too much. Even so, a $10 bottle is normally just fine. Sometimes I'll splurge on a $20 bottle if a friend recommends it. I'm no connoisseur, but I can remember tasting wines in that price range that were as delicious as I can imagine wine tasting to my uneducated palate. Granted, my memories probably have a lot to do with the food we were eating, the friends I was with, the sun going down over the hills in the distance, the smell of woodsmoke drifting up from the fire. In any case, there is no wine in the world that could taste twice that good to me. Not for $40. Not for $4,000. Similarly, drinking twice as much certainly doesn't double my enjoyment of the wine. And to the extent that my memory of that bottle of Rioja is enhanced by the context, that just reinforces my point. It wasn't really about the wine at all. It was about the experience. The quality of most things has an upper limit, which is normally reached rather quickly. If not, what you're seeking probably has less to do with the product in question than with some psychological itch you've been convinced that product can scratch. A watch tells the time; a $20,000 Rolex tells people you've got issues.

In *Stumbling on Happiness*, Daniel Gilbert explains why our species is so easily tricked by carrots dangling just out of reach: "The human brain mispredicts the sources of its own satisfaction, and the reason is that we fail to understand how quickly we will adapt to both positive and negative events. People are consistently surprised by how quickly the abnormal becomes normal, the extraordinary becomes ordinary. When people say, 'I could never get used to that,' they are almost always wrong." This process of quickly taking comforts for granted is known to psychologists as "hedonic adaptation," and it undermines our struggle for happiness by leading us to misplace our energy in pursuit of initially novel states that quickly become normal—addiction, in other words.

The best shower I ever took was in 1987, in Nepal, where I'd been walking in the mountains for several dusty days. When I made it to camp that night, I heated a few liters of water over a small fire. It took a long time to get warm. After carefully sponging targeted areas, I lifted the pot of steaming water over my squatting, shivering body and poured it—slowly and deliciously—over my head and neck. I'll never forget the feeling as it flowed down my spine, warm as blood. Yet I've already forgotten the perfectly hot shower I took this morning. It required nothing more from me than pushing a lever and taking a step into the steaming stream of mind-numbing, totally uninteresting comfort.

In addition to our species' self-defeating tendency to quickly take for granted whatever improved conditions we encounter or create, we're susceptible to external cues telling us where our baselines should be located. In a column called "Downsizing Supersize," economics journalist James Surowiecki points to a study in which "researchers put a bowl of M&M's on the concierge desk of an apartment building, with a scoop attached and a sign below that said 'Eat Your Fill.' On alternating days, the experimenters changed the size of the scoop—from a tablespoon to a quarter-cup scoop, which was four times as big." If people were only eating what they wanted, the scoop size shouldn't have mattered, but it did—a lot. Bigger scoop, more candy. Surowiecki's conclusion: "Most of us don't have a fixed idea of how much we want; instead, we look to outside cues—like the size of a package or cup—to instruct us." And the cues, in American society especially, all point toward *more*.

In Spain, where I lived for many years, the standard pour for a draft beer, known as *una caña*, is 25 centiliters, whereas in the United States, "a beer" in a bar will generally be a pint (47 centiliters). So in Spain, when I go out for a few beers with a friend, my "three beers" will normally amount to 75 centiliters (25 ounces), but in the United States, I'll have imbibed almost twice that amount. No wonder I gain weight when I'm in the States! In my head, the situations are identical: I'm drinking a few beers. But my liver and waistline know better.

Itamar Simonson and Amos Tversky have studied "context-dependent preferences." They showed that if you presented potential customers with a standard inexpensive camera and a more expensive one with more features, about half would go for each. But when an even more expensive third option was added to the mix most people now opted for the middle option. Suddenly, just by adding the possibility of extreme extravagance to the mix, what had previously seemed pricey to many buyers became the reasonable choice. From a cramped seat in coach, business class looks like the promised land. But from your business-class seat, you can hear the tinkling of champagne glasses in first class.

– Rich Asshole Syndrome (RAS) –

In 2007, Gary Rivlin wrote a *New York Times* feature profile of highly successful people in Silicon Valley. One of them, Hal Steger, lived with his wife in a million-dollar house overlooking the Pacific Ocean. Their net worth was about $3.5 million. Assuming a reasonable return of 5 percent, Steger and his wife were positioned to cash out, invest their capital, and glide through the rest of their lives on a passive income of around $175,000 per year after glorious year. Instead, Rivlin wrote, "Most mornings, [Steger] can be found at his desk by 7. He typically works 12 hours a day and logs an extra 10 hours over the weekend." Steger, fifty-one at the time, was aware of the irony (sort of): "I know people looking in from the outside will ask why someone like me keeps working so hard," he told Rivlin. "But a few million doesn't go as far as it used to."

Steger was presumably referring to the corrosive effects of inflation on the currency, but he appeared to be unaware of how wealth was affecting his own psyche. "Silicon Valley is thick with those who might be called working-class millionaires," wrote Rivlin, "nose-to-the-grindstone people like Mr. Steger who, much to their surprise, are still working as hard as ever even as they find themselves among the fortunate few. But many such accomplished and ambitious members of the digital elite still do not think of themselves as particularly fortunate, in part because they are surrounded by people with more wealth—often a lot more."

After interviewing a sample of executives for his piece, Rivlin concluded that "those with a few million dollars often see their accumulated wealth as puny, a reflection of their modest status in the new Gilded Age, when hundreds of thousands of people have accumulated much vaster fortunes." Gary Kremen was another glaring example. With a net worth of around $10 million as the founder of Match.com, Kremen understood the trap he was in, but still he wasn't ready to let go of the mango: "Everyone around here looks at the people above them," he said. "You're nobody here at $10 million." If you're nobody with $10 million, what's it cost to be somebody?

Now, you may be thinking, "Fuck those guys and the private jets they rode in on." Fair enough. But here's the thing: Those guys are already fucked. Really. They worked like hell to get where they are—and they've got access to more wealth than 99.999 percent of the human beings who have ever lived—but they're still not where they think they need to be. Without a fundamental change in the way they approach their lives, they'll never reach their ever-receding goals. And if the futility of their situation ever dawns on them like a dark sunrise, they're unlikely to receive a lot of sympathy from their friends and family. "At this point, nobody gives a damn what my problem is," explained world-famous millionaire comic Jim Carrey. "I could literally have a tumor on the side of my head and they'd be like, 'Yeah, big deal. I'd *eat* a tumor every morning for the kinda money you're pulling down.'"

The Spanish word *aislar* means both "to insulate" and "to isolate," which is what most of us do when we get more money. We buy a car so we can stop taking the bus. We move out of the apartment with all those noisy neighbors into a house behind a wall. We stay in expensive, quiet hotels rather than the funky guest-houses we used to frequent. We use money to insulate ourselves from the risk, noise, inconvenience. But the insulation comes at the price of isolation. Our comfort requires that we cut ourselves off from chance encounters, new music, unfamiliar laughter, fresh air, and random interaction with strangers.

Researchers have concluded again and again that the single most reliable predictor of happiness is feeling embedded in a community. In the 1920s, around 5 percent of Americans lived alone. Today, more than a quarter do—the highest levels ever, according to the Census Bureau. Meanwhile, the use of antidepressants has increased over 400 percent in just the past twenty years and abuse of pain medication is a growing epidemic. Correlation doesn't prove causation, but those trends aren't unrelated. Maybe it's time to ask some impertinent questions about formerly unquestionable aspirations, such as comfort, wealth, and power.

My first real job after college—if you don't count gutting salmon in Kenai, Alaska, for a few weeks in the summer of '84—was commercial real estate management in New York's Diamond District. Armed with my BA in English and my fishy work experience, a gig negotiating leases with Hasidic diamond dealers in Midtown Manhattan was about as likely a career move for me as bullfighting or ballet.

The patriarch of the family who owned the buildings I was hired to help manage was in his early seventies, and very, very wealthy. To say he was set for life is like saying Lake Huron is unlikely to run dry. In addition to the real estate, he had interests in several other businesses—one of which involved extracting precious metals from hearing aid batteries. But not all the batteries contained platinum—or whatever it was they were after. So while I dealt with the gold dealers, plumbers, city inspectors, and gemstone cutters, my multimillionaire boss came to his windowless office every morning and dumped hundreds of tiny batteries from a big Mason jar onto his big, oak desk. Coffee at hand, he spent his mornings sorting the valuable ones from the others. After lunch, a Cuban cigar and glass of Glenfiddich Scotch replaced the coffee, but the sorting continued. Every day, Monday through Friday.

One day I asked him why he didn't go somewhere and enjoy his money—go to his place in Jamaica, travel to Europe, whatever. "Money has no real value for me anymore, Chris," he said. "It's like points in a game. And I like winning." But if "winning" means getting up every morning to shave, put on a suit and tie, and commute to a windowless office where you sit alone sorting hearing aid batteries into two piles, what kind of game are we playing?

Not long after quitting that strange job in Manhattan, I realized that I, too, was rich. I'd been traveling for a few months in India, ignoring the beggars as best I could. Having lived in New York, I was accustomed to averting my attention from desperate adults and psychotics, but I was having trouble getting used to the groups of children who would gather right next to my table at street-level restaurants, staring hungrily at the food on my plate. Eventually, a waiter would come and shoo them away, but they'd just run out to the street and watch from there—waiting for me to leave the waiter's protection, hoping I'd bring some scraps with me.

In New York, I'd developed psychological defenses against the desperation I saw on the streets. I told myself that there were social services for homeless people, that they would just use my money to buy drugs or booze, that they'd probably brought their situation on themselves. But none of that worked with these Indian kids. There were no shelters waiting to receive them. I saw them sleeping on the streets at night, huddled together for warmth, like puppies. They weren't going to spend my money unwisely. They weren't even asking for money. They were just staring at my food like the starving creatures they were. And their emaciated bodies were brutally clear proof that they weren't faking their hunger.

A few times, I bought a dozen samosas and handed them out, but the food was gone in an instant, and I was left with an even bigger crowd of kids (and, often, adults) surrounding me with their hands out, touching me, seeking my eyes, pleading. I knew the numbers. With what I'd spent on my one-way ticket from New York to New Delhi, I could have pulled a few families out of the debt that would hold them down for generations. With what I'd spent in New York restaurants the year before, I could have put a few of those kids through school. Hell, with what I'd budgeted for a year of traveling in Asia, I probably could have *built* a school.

I wish I could tell you I did some of that, but I didn't. Instead, I developed the psychological scar tissue necessary to ignore the situation. I learned to stop thinking about things I could have done but knew I wouldn't. I stopped making facial expressions that suggested I had any capacity for compassion. I learned to step over bodies in the street—dead or sleeping—without looking down. I learned to do these things because I had to—or so I told myself.

Research conducted at the University of Toronto by Stéphane Côté and colleagues confirmed that the rich are less generous than the poor, but their findings suggest it's more complicated than simply wealth making people stingy. Rather, it's *the distance created by wealth differentials* that seems to break the natural flow of human kindness. Côté found that "higher-income individuals are only less generous if they reside in a highly unequal area or when inequality is experimentally portrayed as relatively high." Rich people were as generous as anyone else when inequality was low. The rich are less generous when inequality is extreme, a finding that challenges the idea that higher-income individuals are just more selfish. If the person who needs help doesn't seem *that* different from us, we'll probably help them out. But if they seem too far away (culturally, economically), we're less likely to lend a hand.

The social distance separating rich and poor, like so many of the other distances that separate us from each other, only entered human experience after the advent of agriculture and the hierarchical civilizations that followed, which is why it's so psychologically difficult to twist your soul into a shape that allows you to ignore starving children standing close enough to smell your curry. You've got to silence the inner voice calling for justice and fairness. But we silence this ancient, insistent voice at great cost to our own psychological well-being.

What if most rich assholes are made, not born? What if the cold-heartedness so often associated with the upper crust isn't the result of having been raised by a parade of resentful nannies, too many sailing lessons, or repeated caviar overdoses, but the compounded disappointment of being lucky but still feeling unfulfilled? We're told that those with the most toys are winning, that money represents points on the scoreboard of life. But what if that tired story is just another facet of a scam in which we're *all* getting ripped off?

Calling the miserable rich "winners" is like calling everyone who ever wore a military uniform a "hero." We're reinforcing the false narrative that generated the mess in the first place. It's true that psychopaths are drawn to lucrative professions, but true psychopaths are rare—even on Wall Street. I'm not saying I'd rather be homeless than wealthy or that there's no substantive difference in life satisfaction between the two situations. But I *am* arguing that being wealthy isn't what it's cracked up to be—not nearly so—and that those who spend their lives chasing wealth that they think will bring them happiness are trapped running on the same wheel as everybody else.

A wealthy friend of mine recently told me, "You get successful by saying 'yes,' but you need to say 'no' a lot to stay successful." If you're perceived to be wealthier than those around you, you'll have to say "no" a lot. You'll be constantly approached with requests, offers, pitches, and pleas—whether you're in a Starbucks in Silicon Valley or the back streets of Calcutta. Refusing sincere requests for help doesn't come naturally to our species. Neuroscientists Jorge Moll, Jordan Grafman, and Frank Krueger of the National Institute of Neurological Disorders and Stroke (NINDS) have used fMRI machines to demonstrate that altruism is deeply embedded in human nature. Their work suggests that the deep satisfaction most people derive from altruistic behavior is not due to a benevolent cultural overlay, but to the evolved architecture of the human brain. When volunteers in their studies placed the interests of others before their own, a primitive part of the brain normally associated with food or sex was activated. When researchers measured vagal tone (an indicator of feeling safe and calm) in seventy-four preschoolers, they found that children who'd donated tokens to help sick kids had much better readings than those who'd kept all their tokens for themselves. Jonas Miller, the lead investigator, said that the findings suggested "we might be wired from a young age to derive a sense of safety from providing care for others." But Miller and his colleagues also found that whatever innate predisposition our species has toward charity is influenced by social cues. Children from wealthier families shared fewer tokens than the children from less-well-off families.

According to Joshua D. Greene, a Harvard neuroscientist and philosopher, many lines of research suggest that morality arises from basic brain activities. Morality, in Greene's view, is not "handed down" by philosophers and clergy, but "handed up," an outgrowth of the brain's basic propensities. When Saint Francis of Assisi said that "it is in giving that we receive," he wasn't pleading a weak case. He was noting a salient characteristic of our species.

Apocalyptic views of human nature are further undercut by research suggesting that the human impulse toward cooperation and other prosocial behaviors has roots that reach far into our prehuman past. "Chimpanzees live in a rich social environment, they depend on each other," says Felix Warneken of Harvard University. "It does not require a big society with social norms to elicit a deep-rooted sense that we care about others." (Frans de Waal and others have demonstrated that our other closest primate cousin, bonobos, are even more deeply cooperative than chimps.) James Rilling uses fMRI to compare brain structure and function in primates (including humans) with the goal of identifying specific brain specializations and deepening our understanding of human brain evolution. He's concluded that human beings have "emotional biases toward cooperation that can only be overcome with effortful cognitive control." Our default behavior is toward cooperation—not raw self-interest.

Researchers working with primates have shown that an ape will go out of her way to give a companion access to food, even if she gets less as a result. When capuchin monkeys are offered two different-colored tokens—one of which rewards only the recipient while the other brings a treat to both monkeys—they develop a preference for the "prosocial" token. De Waal explains, "This is not out of fear, because dominant monkeys (who have least to fear) are in fact the most generous."

With monkeys, as with humans, generosity comes together with an expectation of fairness. In experiments de Waal ran with Sarah Brosnan, monkeys got a slice of cucumber or a grape for doing a task. The monkeys were fine as long as they were getting the same "payment," whether it was high (a grape) or low (a cucumber slice). But when the researchers introduced unequal pay into the experiment, things got tense. "The monkey receiving cucumber contentedly munches on her first slice, yet throws a tantrum after she notices that her companion is getting grapes," reported de Waal.

Interestingly, just handing out unequal foods doesn't prompt the same kind of response in the primates. The foods need to be given in exchange for some kind of task for the fairness response to be triggered. When Brosnan conducted similar studies with chimpanzees, she observed "second-order fairness," where even the winners balked at the arrangement. "We unexpectedly found that chimpanzees were more likely to refuse a high-value grape when the other chimpanzee got a lower-value carrot than when the other chimpanzee also received a grape."

– Drunk on Dollars –

Psychologists Dacher Keltner and Paul Piff monitored intersections with four-way stop signs and found that people in expensive cars were four times more likely to cut in front of other drivers, compared to folks in more modest vehicles. When the researchers posed as pedestrians waiting to cross a street, all the drivers in cheap cars respected their right of way, while those in expensive cars drove right on by 46.2 percent of the time, even when they'd made eye contact with the pedestrians waiting to cross. Other studies by the same team showed that wealthier subjects were more likely to cheat at an array of tasks and games. For example, Keltner reported that wealthier subjects were far more likely to claim they'd won a computer game—even though the game was rigged so that winning was impossible. Wealthy subjects were more likely to lie in negotiations and excuse unethical behavior at work, such as lying to clients in order to make more money. When Keltner and Piff left a jar of candy in the entrance to their lab with a sign saying whatever was left over would be given to kids at a nearby school, they found that wealthier people stole more candy from the babies.

Researchers at the New York State Psychiatric Institute surveyed forty-three thousand people and found that the rich were far more likely to walk out of a store with merchandise they hadn't paid for than were poorer people. Findings like this (and the behavior of drivers at intersections) could reflect the fact that wealthy people worry less about potential legal repercussions. If you know you can afford bail and a good lawyer, running a red light now and then or swiping a Snickers bar may seem less risky. But the selfishness goes deeper than such considerations. A coalition of nonprofit organizations called the Independent Sector found that, on average, people with incomes below $25,000 per year typically gave away a little over 4 percent of their income, while those earning more than $150,000 donated only 2.7 percent (despite tax benefits the rich can get from charitable giving that are unavailable to someone making much less).

There is reason to believe that blindness to the suffering of others is a psychological adaptation to the discomfort caused by extreme wealth disparities. Michael W. Kraus and colleagues found that people of higher socioeconomic status were actually less able to read emotions in other people's faces. It wasn't that they cared less what those faces were communicating; they were simply blind to the cues. And Keely Muscatell, a neuroscientist at UCLA, found that wealthy people's brains showed far less activity than the brains of poor people when they looked at photos of children with cancer.

Books such as *Snakes in Suits: When Psychopaths Go to Work* and *The Psychopath Test* argue that many traits characteristic of psychopaths are celebrated in business: ruthlessness, a convenient absence of social conscience, a single-minded focus on "success." But while psychopaths may be ideally suited to some of the most lucrative professions, I'm arguing something different here. It's not just that heartless people are more likely to become rich. I'm saying that being rich tends to corrode whatever heart you've got left. I'm suggesting, in other words, that it's likely the wealthy subjects who participated in Muscatell's study *learned* to be less unsettled by the photos of sick kids *by the experience of being rich*—much as I learned to ignore starving children in Rajastan so I could comfortably continue my vacation.

In an essay called "Extreme Wealth Is Bad for Everyone—Especially the Wealthy," Michael Lewis observed, "It is beginning to seem that the problem isn't that the kind of people who wind up on the pleasant side of inequality suffer from some moral disability that gives them a market edge. The problem is caused by the inequality itself: it triggers a chemical reaction in the privileged few. It tilts their brains. It causes them to be less likely to care about anyone but themselves or to experience the moral sentiments needed to be a decent citizen."

Sukhvinder Obhi, a neuroscientist at Wilfrid Laurier University in Ontario, Canada, wanted to understand how power affects cerebral functioning. Along with his colleagues Jeremy Hogeveen and Michael Inzlicht, Obhi randomly assigned to subjects a feeling of being powerful or powerless by asking them to write about a time they were either dependent on others for help or in absolute control of a situation involving others. Then the subjects watched an incredibly boring video of a hand squeezing a rubber ball, while the scientists monitored the activity of mirror neurons in the subjects' brains. Mirror neurons are key to human compassion; they fire whether you are skiing down a mountain or watching someone else ski down a mountain. The mirror system is the part of the brain that allows us to get inside each other's heads. What Obhi and his colleagues found helps explain why poor people give away a greater proportion of what they have than rich people do: *powerlessness boosts the mirror system, but power dampens it.* Dacher Keltner (the guy who studied assholes in BMWs blowing by old ladies waiting to cross the street) agrees: "Power diminishes all varieties of empathy." Ultimately, diminished empathy is self-destructive. It leads to social isolation, which is strongly associated with sharply increased health risks, including stroke, heart disease, depression, and dementia.

In one of my favorite studies, Keltner and Piff decided to tweak a game of Monopoly.* The psychologists rigged the game so that one player had huge advantages over the other from the start. They ran the study with over a hundred pairs of subjects, all of whom were brought into the lab where a coin was flipped to determine who'd be "rich" and "poor" in the game. The randomly chosen

---

* Monopoly was originally called The Landlord's Game, and was invented as a teaching tool used to demonstrate the evils of concentrated ownership and the tendency of wealth to accumulate in the hands of the already rich.

"rich" player started out with twice as much money, collected twice as much every time they went around the board, and got to roll two dice instead of one. None of these advantages was hidden from the players. Both were well aware of how unfair the situation was. Still, the "winning" players showed the telltale symptoms of Rich Asshole Syndrome. They were far more likely to display dominant behaviors like smacking the board with their piece, loudly celebrating their superior skill, even eating more pretzels from a bowl positioned nearby.

After fifteen minutes, the experimenters asked the subjects to discuss their experience of playing the game. When the rich players talked about why they'd won, they focused on their brilliant strategies rather than the fact that the whole game was rigged to make it nearly impossible for them to lose. "What we've been finding across dozens of studies and thousands of participants across this country," said Piff, "is that as a person's levels of wealth increase, their feelings of compassion and empathy go down, and their feelings of entitlement, of deservingness, and their ideology of self-interest increases."

Of course, there are exceptions to these tendencies. Some wealthy people have the wisdom to navigate the difficult currents their good fortune generates without succumbing to RAS—but such people are rare, and tend to come from humble origins. Perhaps an understanding of the debilitating effects of wealth explains why some who have built large fortunes are vowing not to pass their wealth to their children. Several billionaires, including Chuck Feeney, Bill Gates, and Warren Buffett, have pledged to give away all or most of their money before they die. Buffett has famously said that he intends to leave his kids "enough to do anything, but not enough to do nothing." The same impulse is expressed among those lower on the millionaire totem pole. According to an article on CNBC.com, Craig Wolfe, the owner of CelebriDucks, the largest custom collectible rubber duck manufacturer, intends to leave the millions he's made to charity, which is amazing—but nowhere near as amazing as the fact that someone made millions of dollars selling *collectible rubber ducks*.

Do you know someone who suffers from RAS? There may be help for them. UC Berkeley researcher Robb Willer and his team conducted studies in which participants were given cash and instructed to play games of varying complexity that would benefit "the public good." Participants who showed the greatest generosity benefited from more respect and cooperation from their peers and had more social influence. "The findings suggest that anyone who acts only in his or her narrow self-interest will be shunned, disrespected, even hated," Willer said. "But those who behave generously with others are held in high esteem by their peers and thus rise in status."

Keltner and Piff have seen the same thing. "We've been finding in our own laboratory research that small psychological interventions, small changes to people's values, small nudges in certain directions, can restore levels of egalitarianism and empathy," said Piff. "For instance, reminding people of the benefits of cooperation, or the advantages of community, cause wealthier individuals to be just as egalitarian as poor people." In one study, they showed subjects a short video—just forty-six seconds long—about childhood poverty. They then checked the subjects' willingness to help a stranger presented to them in the lab who appeared to be in distress. An hour after watching the video, rich people were as willing to lend a hand as were poor subjects. Piff believes these results suggest that "these differences are not innate or categorical, but are malleable to slight changes in people's values, and little nudges of compassion and bumps of empathy."

Piff's conclusions align with the lessons passed along by thousands of generations of our foraging ancestors, whose survival depended on developing social webs of mutual aid. Selfishness, they understood, leads only to death: first social, and then ultimately, biological. While the neo-Hobbesians struggle to explain how human altruism can exist, other scientists question their premise, asking if there's any functional utility to selfishness. "Given how much is to be gained through generosity," says Robb Willer, "social scientists increasingly wonder less why people are ever generous and more why they are ever selfish."

Decades of "greed is good" messaging has sought to remove a sense of shame from being a beneficiary of outrageous extremes of wealth inequality. Still, the shame lingers, because the messaging runs up against one of our species' deepest innate values. Institutions seeking to justify a fundamentally antihuman economic system constantly rebroadcast the message that winning the money game will bring satisfaction and happiness. But we've got around three hundred thousand years of ancestral experience telling us it just isn't so. Selfishness may be essential to civilization, but that only raises the question of whether a civilization so out of step with our evolved predispositions makes sense for the human beings within it.

Part IV

A PREHISTORIC PATH
INTO THE FUTURE

———

We call ourselves, somewhat presumptuously, *Homo sapiens sapiens*: the hominid that knows it knows. But what, exactly, do we know we know that no other creature knows? We know we will suffer and we know we will die—and this knowledge can drive us to distraction. In order to move toward a future worthy of our origins, we will have to turn and face the fears we've been running from since first stepping onto the spinning wheel of civilization.

Chapter 10

All's Well That Ends Well

Perhaps the whole root of our trouble, the human trouble, is that we will sacrifice all the beauty of our lives, will imprison ourselves in totems, taboos, crosses, blood sacrifices, steeples, mosques, races, armies, flags, nations, in order to deny the fact of death, the only fact we have.

—James Baldwin

All goes onward and outward—nothing collapses; And to die is different from what any one supposed, and luckier.

—Walt Whitman

In *Civilization and Its Discontents*, Freud attributed the chronic psychopathology of the civilized to the suppression of instinctive, primarily sexual, urges. Writing in the age of steam engines, it's not surprising that Freud saw civilization in terms of primordial urges contained, pressurized, and redirected away from their natural, immediate release—harnessed for more productive results. For Freud, civilization is the result of pleasure denied or, at least, delayed and deflected.

No doubt there's a lot of that going on, but from my perspective, the pyramids, the cathedrals, the Pentagon, Wall Street, and the Great Wall of China are also expressions of another kind of civilizational hysteria generated by a refusal to accept *the* insight that defines us as human beings. In *The Denial of Death*, Ernest Becker wrote, "The idea of death, the fear of it, haunts the human animal like nothing else; it is a mainspring of human activity—activity designed largely to avoid the fatality of death, to overcome it by denying in some way that it is the final destiny for man." But like the night, death is inevitable. We've only succeeded in breaking it into innumerable fragmentary shadows that dim the day. In our panic to avoid the darkness of death, we sacrifice the light of our lives.

The psychological underpinnings of Becker's insights have been explored in experiments by Sheldon Solomon, Jeff Greenberg, and Thomas Pyszczynski, who have spent decades investigating the ways by which we subconsciously seek to deny our mortality by aligning our personal identity with totems, taboos, religions, and armies. Culture, they believe, offers a refuge from existential terror by providing meaning and guidance. If we follow the rules, we may even have hope of immortality in the form of an afterlife or reincarnation, or it may be symbolic: monuments, works of art, or streets bearing our names.

Dozens of experiments have shown how our behavior changes when we are subtly reminded of our mortality. In their first study, Solomon and his colleagues asked a group of municipal judges to think of something unpleasant, while another group was indirectly reminded that they would die one day. Then both groups were asked to set bond for various alleged crimes. Judges who had been reminded of their own eventual demise set bond amounts nine times higher. Solomon and his colleagues have repeatedly demonstrated that reminders of mortality make people far more critical of those outside their belief system while clinging more closely to those within it.

Sometimes, our pathetic gestures toward immortality extend even into the grave. Caitlin Doughty, the famed "hip mortician" and author of *Smoke Gets in Your Eyes*, told me about a horribly funny, unintended consequence of the strange American obsession with high-tech caskets. It seems that these hermetically sealed, stainless-steel, death-denying capsules sometimes explode, due to the pressurized gases created by the decomposing corpse inside. What do we think we're accomplishing by pumping dead bodies full of formaldehyde and sealing them up in wormproof caskets? Is the $10,000 casket a symbolic last barrier against death? Or is the steel box meant to protect the living from what we've locked inside and buried in the ground, like radioactive waste?

Why worry about death? It's the dying that keeps *me* awake at night. It's a simple distinction, yet the two concepts have a terrible tendency to seep into each other. When the game is over, it's over, and the lights go out. This—whatever it is—will be no more. Fearing death is literally being afraid of nothing at all. Yet civilization amplifies and is fueled by our fear of death, when it's the dying we should really be concerned with.

The tragic irony of our situation is that our medical advances, our sterile surgical technology, pharmaceutical wonders, and refined medical procedures have proven worse than useless when it comes to easing the agonies of dying. As Susan Jacoby explains in *Never Say Die: The Myth and Marketing of the New Old Age*, while death remains what it's always been, dying is getting harder: "If a decent death is defined by the absence of extended suffering, an American who lives into advanced old age in the twenty-first century probably has less chance of receiving that mercy than the poorest peasant did in the fourteenth century." A study published in the *Annals of Internal Medicine*, in 2015, in which 7,204 patients were monitored, confirmed Jacoby's sense that the situation is getting worse in the United States. Despite all our efforts to improve end-of-life care, reports of pain in the last year of life actually went up from 1998 to 2010. Researchers found that the prevalence of pain had increased 11.9 percent and that reports of depression and periodic confusion were up more than 26 percent. One of the study's authors, Joanne

Lynn, suspects that medical advances are partly responsible for the increases in patient agony. "Maybe we've made more medical stuff coming at people that maybe lets them live a little bit longer, but under much more burdensome circumstances," she said. "You're still going to have to find some way off this terrestrial globe," she said, "and it may as well be as comfortable, meaningful, dignified and inexpensive as it can possibly be."

Comfortable, meaningful, dignified, and *inexpensive*? Faced with end-of-life issues, many recoil at practical economic considerations. This predictable, generous, humane impulse to ignore costs when confronted with existential matters is based on the assumption that spending more will benefit the patient, when the relationship between expense and outcome is often perversely the inverse. Although the United States spends more on health care than any other country in the world—more than $9,000 per capita in 2013—it ranks dead last among advanced countries on a variety of health measures. Rather than prolonging life, we appear to be extending the process of dying. "For all its technological sophistication and hefty price tag," wrote internist Craig Bowron in the *Washington Post*, "modern medicine may be doing more to complicate the end of life than to prolong or improve it. With unrealistic expectations of our ability to prolong life, with death as an unfamiliar and unnatural event, and without a realistic, tactile sense of how much a worn-out elderly patient is suffering, it's easy for patients and families to keep insisting on more tests, more medications, more procedures."

And of course, it's not just well-intentioned, misinformed families. Doctors and medical facilities are often responding to perverse financial incentives that reward them for performing expensive, painful procedures even when they are of no benefit to the patient. About 30 percent of all Medicare expenditures are attributed to the 5 percent of beneficiaries who die each year, with a third of that money being spent in the last month of the patients' lives. How much should it cost to help a dying person go in peace? "At a certain stage of life," concludes Bowron, "aggressive medical treatment can become sanctioned torture."

Atul Gawande, a surgeon and author of several books about his experiences in medicine, comes to similar conclusions in *Being Mortal: Medicine and What Matters in the End.* "Over and over," writes Gawande, "we in medicine inflict deep gouges at the end of people's lives and then stand oblivious to the harm done." Gawande sees that much of this inadvertent cruelty is the result of our unwillingness to confront the fact of mortality head-on: "Our most cruel failure in how we treat the sick and aged is the failure to recognize that they have priorities beyond merely being safe and living longer."

Gawande is not unique in his conflicts over how his profession treats the dying. As it turns out, doctors—the field marshals leading us into this endless, hopeless assault on death—face their own final days differently from what they advise the rest of us to do. Take a look at this chart of how doctors responded when asked if they wanted various common end-of-life interventions.

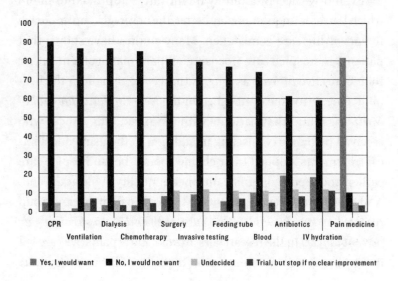

Physicians are reluctant to receive some common procedures because they know what really happens behind the hype. Take CPR, for example. A recent study that looked at how CPR is portrayed on television found that it was shown to be successful 75 percent of the time and that 67 percent of the patients were shown going home. But, in reality, among actual patients who received the intervention, most studies conclude that only 8 percent survived for more than a month afterward and, of these, only 3 percent returned to anything like a normal life. Dr. Ken Murray, clinical assistant professor of family medicine at USC, wrote about his own experience with the procedure: "Many people think of CPR as a reliable lifesaver when, in fact, the results are usually poor. I've had hundreds of people brought to me in the emergency room after getting CPR. Exactly one, a healthy man who'd had no heart troubles . . . walked out of the hospital. If a patient suffers from severe illness, old age, or a terminal disease, the odds of a good outcome from CPR are infinitesimal, while the odds of suffering are overwhelming."

Even if we accept a purely quantitative approach to life in which longer is unquestionably better, the never-give-up approach to end-of-life care is mistaken. Many studies have shown that patients in hospice care (focusing on pain management rather than cure) live *at least* as long as those who remain in the hospital. One 2010 study published in the *New England Journal of Medicine* found that patients with advanced lung cancer who received palliative counseling in addition to the usual oncological treatments stopped their chemo sooner, began hospice care earlier, and reported significantly better quality of life at the end. And even though fewer patients in the early palliative care group received aggressive end-of-life care, *they ended up surviving 25 percent longer*. In the researchers' words, "Early palliative care led to significant improvements in both quality of life and mood. As compared with patients receiving standard care, patients receiving early palliative care had less aggressive care at the end of life but longer survival."

Despite studies showing how more compassionate, realistic approaches to end-of-life care result in far less suffering at substantially lower cost, the American medical and political establishments largely continue to ignore the real needs of dying people. Medicare will pay for expensive surgery to keep an ill ninety-year-old alive for a few more painful months, but refuses to pay for much cheaper home care that could keep the same person out of medical institutions. "You can't believe the forces of the system that are arrayed against [being allowed to die at home]," said Jack Resnick, a former health system executive, currently working as a physician in New York. "These decisions are not made on the basis of what the individuals need. They're based on what the institutions need."

When investigative journalist Katy Butler's father suffered a stroke at the age of seventy-nine, she suddenly found herself in the imperfect storm of the American health-care system. She eventually published a book about what she learned, called *Knocking on Heaven's Door: The Path to a Better Way of Death*. She found that Medicare won't pay doctors for the time it takes to offer sound advice, but it will pay for expensive drugs and devices: "The corporate healthcare lobbies help determine what doctors get paid to do. We pay doctors very well for deploying technology and very poorly for spending time with patients. This shapes their behavior."

One would think that a country with chronic budget deficits and a huge segment of the population about to enter their final years would enthusiastically encourage conversations about redirecting medical spending in ways that both increase quality of life and reduce costs. But so far, at least, that's not happening. Here we have a situation where we really can do far more with far less. A recent study in the *Archives of Internal Medicine* found that the less money is spent in the final weeks of life, the better things go for the patient. But these monetary savings required physicians to have honest end-of-life (EOL) conversations with their patients, in which they speak frankly about the situation rather than pitching some miraculous drug or surgical procedure they know is expensive, painful, and probably futile. In the stark language of the study's conclusions: "Advanced cancer patients who reported EOL conversations with physicians had significantly lower health care costs in their final weeks of life. Higher costs were associated with worse quality of death."

Difficult as it is to acknowledge, sometimes there's no point in raging against the dying of the light. Sometimes, giving up with grace and dignity is the right thing to do. "We have a culture that has bought into the idea that medicine is supposed to save your life," says Daniel Callahan, research scholar and president emeritus at the Hastings Center, a nonpartisan bioethics research center. "But no matter how [many] medical treatments we get, it's never good enough because people eventually die. You can save them from one thing, but then death gets [them] one way or another. We're not in a winnable war against death."

Foragers, living in the immediate presence of death, understand that the end is eventually unavoidable. In *The World Until Yesterday*, Jared Diamond delineates the five most common ways elderly or terminally ill foragers make the transition. In some societies, they are simply neglected until they die. Others abandon the dying person when the group moves on from one camp to another. Some, including the Inuit, Crow, and Yakut people, encourage the dying person to take his or her own life by drifting out to sea or jumping from a cliff. A more active approach is to assist the "voluntary" suicide by choking or a blow to the back of the head. And, finally, the same approach is sometimes taken without the victim's knowledge or consent, when the person can no longer keep up with the group or contribute to the general welfare.

While these primitive ways of hastening death will no doubt strike many of us as barbaric, are they really worse than our "civilized" approaches to death? Millions of elderly people are sent away to what amount to foundling hospitals for the aged: a place to be neglected until they die. In the United States and most other Western nations, even terminally ill people face furious resistance to being allowed to step through their final door with dignity and intention intact. Only a handful of states allow doctor-assisted euthanasia, and even there, the regulations are onerous. If we don't own our lives, what, then, is ours?

Not much has changed concerning how American law handles compassionate death since the case of *Gilbert* v. *State of Florida* in 1986. Roswell Gilbert had to make excruciating decisions about how best to help his wife, Emily. After years of constant, debilitating pain from severe arthritis and osteoporosis as well as advanced Alzheimer's disease, Emily was rejected by three nursing homes and the local hospital—none of which could handle the suffering, delusional woman. Emily had stopped eating, she suffered unrelenting pain from multiple bone fractures due to her osteoporosis, and there was no hope of recovery. Her last words were, "Ros, I love you dearly. God, I want to die." Her husband of fifty-one years shot her twice in the head. He later said he felt grief, but no regret. "I stood there and cried thinking my wife was dead. But the fact that she was no longer suffering gave me relief."

If Emily had been a family pet, we'd congratulate Ros for his courage and mercy, but such mercy apparently can't be shown to the human beings we love, just to our cats and dogs. Roswell Gilbert, at seventy-five years of age, was convicted of first-degree murder and sentenced to twenty-five years in prison.

I don't mean to minimize the complexities of situations like that faced by Ros and Emily. There are far better ways to handle such horrible suffering than a desperate gunshot to the head, but Ros and Emily were forced into their fate by a medical system reflective of a culture that insists that death is a failure rather than an inevitable element of life. We often hear things like, "He lost a long battle with cancer" and "She's going to fight this thing." But we rarely hear, "He sensibly decided not to bother with months of chemotherapy in favor of a few pain-free weeks with his family" or "She bravely opted to end her life peacefully now rather than burden her family with further medical bills they can't afford for the possibility of a few more months of agony." And while debate rages over whether a fetus qualifies as a human being with legal rights, an equally urgent question, not often asked, is whether someone in her nineties, with Alzheimer's disease so advanced that she cannot recognize her closest friends or family, feed herself, or even remember her own name, is still among the living. If not, is she an appropriate recipient of life-extending procedures that might better be provided to patients with decades of life ahead of them?

One of the most prominent physicians in the country, Ezekiel J. Emanuel, published a bombshell article in *The Atlantic* in 2014 called "Why I Hope to Die at 75." The article's subtitle sums up Emanuel's position: "Society and families—and you—will be better off if nature takes its course swiftly and promptly." An oncologist and bioethicist, Emanuel acknowledges that death is a loss. But, he writes, "Living too long is also a loss. It renders many of us, if not disabled, then faltering and declining, a state that may not be worse than death but is nonetheless deprived. It robs us of our creativity and ability to contribute to work, society, the world. It transforms how people experience us, relate to us, and, most important, remember us."

Some ways of living *are* worse than death. A recent paper with the chilling title "States Worse Than Death Among Hospitalized Patients With Serious Illnesses" concludes that "studies among healthy outpatients and those with serious illnesses show that a significant minority, and sometimes a majority, rate states such as severe dementia as worse than death." And let's also consider how these states affect the families of those who are trapped by a health-care system mindlessly opposed to death at any cost.

Dr. Emanuel (fifty-seven at the time of writing his article) isn't planning to commit suicide on his seventy-fifth birthday—and is opposed to legalizing physician-assisted suicide. He's not rejecting the idea of living beyond seventy-five, but he is against what he calls "the American immortal." He views our "desperation to endlessly extend life" as "misguided and potentially destructive." The years added to the end of our lives aren't typically healthy and active. In the past few decades, Emanuel writes, "increases in longevity seem to have been accompanied by increases in disability—not decreases." He points to research showing that in 1998 about 28 percent of American men over eighty had functional limitation, but by 2007 that number had increased to 42 percent. For women, the results were even worse, with more than half the women over eighty facing significant limitations on their ability to function independently. Eileen Crimmins, who conducted this research, concluded that the much-celebrated increase in overall longevity was actually a net loss in terms of *functional longevity*: "Increase in the life expectancy with disease and a decrease in the years without disease. The same is true for functioning loss, an increase in expected years unable to function." It seems we haven't prolonged our lives much after all. We've merely made our own suffering slow-motion.

Chapter 11

In the Absence of the Sacred

Whether . . . civilization has most promoted or most injured
the general happiness of man is a question that may be strongly
contested. [Both] the most affluent and the most miserable of
the human race are to be found in the countries that are called
civilized.

—Thomas Paine, *Agrarian Justice* (1795)

"Don't Worry, Be Happy" may be the most annoying song ever
written—the musical equivalent of someone telling you to smile.
I suspect that Viktor Frankl—neurologist, psychiatrist, and Holocaust survivor—would be even more annoyed than I am by Bobby
McFerrin's empty tropical optimism. Frankl believed that a sense
of *meaning*, not happiness, is the essential ingredient in a worthwhile life. Happiness, when it comes, is an incidental bonus, in
Frankl's view—an occasional gift that should never be sought and
that soon slips through our grasping fingers anyway. The pursuit
of happiness, he believed, just ends up leaving us unhappy about
having failed to capture it—thus compounding the problem we
started with.

The experience of foragers lends little support to the a priori
Hobbesian assumption underlying Frankl's existentialism (and
Buddhist teachings and Christian dogma) holding that the basic
human state is miserable. This assumption is foundational to a
great deal of both popular and philosophical conjecture. A 2006

article by John Lanchester in *The New Yorker* called "Pursuing Happiness" begins with the utterly unsubstantiated yet widely believed assertion that our foraging ancestors "would have regarded our easy, long, riskless lives with incredulous envy." According to Lanchester, our ancestors "would have regarded us as so lucky that questions about our state of mind wouldn't be worth asking."

Really? I suspect they'd have plenty of questions about our state of mind, starting with, "Why are so many people so lonely? Why is war constant? Why are so many of you living in such pain that you drug yourselves, often, to death? Why is it necessary to keep millions of people in prisons?" The notion that things are so great today that precivilized humans would be dumbfounded at how lucky we are is undermined by research like that published in the *Journal of Affective Disorders* in 2012, which warned, "The economic and marketing forces of modern society have engineered an environment . . . that maximize[s] consumption at the long-term cost of well-being. In effect," the authors concluded, "humans have dragged a body with a long hominid history into an overfed, malnourished, sedentary, sunlight-deficient, sleep-deprived, competitive, inequitable, and socially-isolating environment with dire consequences." Hardly the modern paradise the neo-Hobbesians keep insisting we're living in.

Recall that the "pervasive happiness" of the Pirahã is due, in Everett's estimation, to their "ability to handle anything that their environment throws at them [so] that they can enjoy whatever comes their way." Similarly, reflecting on her years observing how the Yequana raise their children, Liedloff pointed to the bedrock satisfaction omnipresent in their lives, but that is only a momentary experience in ours, where *happiness ceases to be a normal condition of being alive, and becomes a goal.* Back in the United States, she was struck by our inability to understand what's missing in our lives: "What was once man's confident expectation of suitable treatment and surroundings is now so frustrated that a person often thinks himself lucky if he is not actually homeless or in pain. But even as he is saying, 'I'm all right,' there is in him a sense of loss, a longing for something he cannot name, a feeling of being off-center, of missing something."

The continuum has been broken because the human animal no longer lives in a human world. We live in a world created by and for institutions that thrive on commerce, not human beings that thrive on community, laughter, and leisure. "The expectations and tendencies of the human species," in Liedloff's words, no longer play out "in an environment consistent with that in which those expectations and tendencies were formed."

It doesn't matter how many times we're told we live in the promised land or, even, how deeply we believe it to be true. The human animal is sickened by the disconnect between the nutrition it evolved to expect and the sugary nonsense it encounters. Even if relentless advertising gets you to believe soft drinks are nutritious, your body knows better, and is likely to respond with tooth decay, diabetes, and heart disease. Even those who believe themselves to be content may not be. "Their perfect adjustment to that abnormal society is a measure of their mental sickness," wrote Aldous Huxley, referring to "millions of abnormally normal people, living without fuss in a society to which, if they were fully human beings, they ought not to be adjusted."

Whether one measures the value of life in the currency of happiness, meaning, interestingness, or merely the absence of despair, the subtle traumas of modern life are inescapable. A 2013 Gallup poll revealed that 70 percent of Americans hate their jobs or have simply "checked out" of them, while only 30 percent are "engaged and enthusiastic" about what they spend forty-plus hours per week doing. As Thoreau noted long ago, "Most men would feel insulted, if it were proposed to employ them in throwing stones over a wall, and then in throwing them back, merely that they might earn their wages. But many are no more worthily employed now."

Not surprisingly, the use of antidepressants in the United States is up nearly 400 percent since 1990. In 2008, 23 percent of women between the ages of forty and fifty-nine were taking at least one of them. In 1985, sociologists asked Americans if they had close friends in whom they could confide. Ten percent said they had no one. By 2004, the number of people so isolated that they had no one they could confide in had risen to 25 percent. The CDC reported in 2013 that the rate of suicide among Americans in the prime of life (from thirty-five to sixty-four years old) had jumped 28.4 percent in the first decade of the twenty-first century, surpassing, for the first time, the number of people who died in car accidents. Among men in their fifties, suicides were up 50 percent, while suicide among women between sixty and sixty-four rose nearly 60 percent.

Given these dire trends, we're fortunate to live in countries with access to modern psychotropic medications, right? Maybe not. As anthropologist T. M. Luhrmann explains, the notion that mental illness is merely a result of brain anomalies misses the subtleties of how human beings interact with our social contexts: "Social experience plays a significant role in who becomes mentally ill, when they fall ill and how their illness unfolds. We should view illness as caused not only by brain deficits but also by abuse, deprivation and inequality, which alter the way brains behave." Many studies have confirmed that children raised in urban environments suffer from more depression and are about twice as likely to suffer from psychosis, as compared to children raised away from cities. When researchers tried to identify precisely what aspects of urban living were responsible for the increased risk of mental illness in children, they concluded that psychotic symptoms were more likely when children lived in neighborhoods with "low social cohesion"—not much interaction between neighbors, low like-lihood that neighbors would intervene to help each other—and when their family had been the victim of a crime. Other studies have pinpointed economic inequality, lack of exposure to nature, and early separation from parents as highly correlated with risk of mental health issues. In other words, the further families are from the kinds of mutual support and social cohesion enjoyed by foragers, the more likely their children are to suffer severe mental illness.

But not all forms of mental illness can be blamed on lone-liness, chronic anxiety, too little exposure to nature, and so on. Schizophrenia, for example, is thought to be highly related to genetics, and to occur at roughly the same rate around the world, independent of culture. But even if the origins of the disorder are not cultural, the likelihood of recovery may be. In the 1970s, the World Health Organization conducted a large-scale study comparing outcomes of people diagnosed with schizophrenia in developing nations with those in the United States and other "advanced" nations. After following patients for five years, the WHO reported that 64 percent of patients in poor countries have "good" outcomes, while less than a third as many (18 percent) in rich countries did as well. The WHO concluded that living in a developed nation was a "strong predictor" that a patient would never fully recover. In response to the predictable uproar these results provoked in the medical community, the WHO followed up with a larger study in the 1980s that attempted to address methodological weaknesses that had been pointed out in the first study (how the patients had been identified, the "developing countries" chosen, what sorts of medications had been used, and so on). The authors of this follow-up study came to similar con-clusions: 63 percent of patients recovered in the poorer countries compared to 37 percent in the "advanced" settings. In the face of further outrage from the medical community, they wrote: "A strong case can be made for a real pervasive influence of a powerful factor which can be referred to as 'culture,' as the context in which gene-environment interactions shape the clinical picture of human disease." Responding to critics of their WHO study, Assen Jablensky and Norman Sartorius noted that "the erosion of social support systems [in developing countries], likely to be associated with the processes of globalization, should be a matter of grave concern. The sobering experience of high rates of chronic disability and dependency associated with schizophrenia in high-income countries, despite access to costly biomedical treatment, suggests that something essential to recovery is missing in the social fabric." It's pretty clear that what we're missing is meaning and one another.

– The Many Voices of God –

> Our greatest blessings come to us by way of madness, provided
> the madness is given to us by divine gift.

> —Socrates

Statistically, you're more likely to hear voices in your head than to be a vegetarian or left-handed. If you do hear voices, before submitting yourself to psychiatric care, you might want to check in with your local chapter of the Hearing Voices Network. An international community of people who experience auditory hallucinations, the HVN was founded in 1988 by the Dutch social psychiatrist Marius Romme, who had a hunch that far more people were hearing voices than generally thought—and that for most of them, it wasn't a problem. There is a subset, of course, for whom the voices are intrusive and deeply disturbing, and Romme noticed that these people had often suffered some form of severe emotional trauma or abuse as children. Romme's insight has been confirmed by massive epidemiological studies such as the Adverse Childhood Experiences (ACE) Study, which looked into eight types of difficult experiences faced by children, ranging from various kinds of physical, sexual, and emotional abuse to having a parent with a mental illness. Having experienced just one of these types of abuse as a child made it 2.5 times more likely that an individual would experience hallucinations later in life. Those unfortunate children who'd experienced seven or all of the different types of trauma were five times more likely to experience disturbing auditory hallucinations than kids who hadn't.

But not all the voices are saying the same things. When Luhrmann compared the reports of voice-hearers in Ghana, the United States, and India, she discovered that while most Americans felt "bombarded" by unfamiliar, hostile voices urging them to harm themselves or others, the Indians and Ghanaians generally believed the voices to be family members or divine figures, who often made helpful suggestions, such as "you should comb your hair" or "it's time to clean up the house."

Luhrmann and her colleagues interviewed sixty adults who'd been diagnosed with schizophrenia: twenty each in San Mateo, California; Chennai, India; and Accra, Ghana. While many of the African and Indian subjects felt that most of their interactions with the voices were positive and even "entertaining," none of the Americans experienced their hallucinations as welcome in any sense. Instead, they tended to view their experiences as evidence of their hopeless sickness. Luhrmann believes her research demonstrates that the harsh, violent voices so common in the West are not an inevitable feature of schizophrenia. If cultural expectations shape the quality and content of auditory hallucinations, "the way people think about thinking changes the way they pay attention to the unusual experiences associated with sleep and awareness, and as a result, people will have different spiritual experiences, as well as different patterns of psychiatric experience." This insight suggests it's time to reassess the current psychiatric tendency to treat the voices heard by people with psychosis as if they are the "uninteresting neurological byproducts of disease which should be ignored," she said.

For our forager ancestors, such voices were anything but "uninteresting neurological byproducts of disease." People who heard them believed they were experiencing a form of divine madness of potentially lifesaving importance and power. A young person who experienced the sorts of hallucinations we associate with severe mental illness would have been seen as a potential shaman—a human being with the ability to move between this world and others. The early manifestations of this ability were terrifying and dangerous, but represented "the call to shamanize." This call cannot be ignored, as the alternative to learning to harness and direct this capacity would be madness or death.

Psychologist Stanley Krippner has spent a lifetime studying how altered states of consciousness (ASC) are used for healing within varying cultural contexts. He notes that researchers have shown that "in 488 societies . . . 89% had one or more forms [of ASC], usually in a ritual or spiritual context. Some were voluntary, such as a shaman's 'journeying' to the 'Upper World,' while others might be partially or completely involuntary, such as a medium's incorporation of a discarnate entity that 'rides' or takes over his or her body, displacing the medium's personality." Krippner concluded that how accommodating cultures are to ASC affects their frequency: "Spontaneous childhood past-life experiences are reported most often in cultural groups marked by beliefs in reincarnation even though a number of cases have been found in Western countries that lack this acceptance. The incidence may be smaller because they occur less frequently or because experients are less likely to share these accounts if they are stigmatized or discounted." Taking this insight to its logical conclusion, Krippner argues that because the capacity to easily enter into altered states would have had such a significant adaptive value—in that it would potentiate placebo and other types of healing activated by states of consciousness—it stands to reason that this capacity would have been selected for in prehistoric populations. In contemporary societies that dismiss such states and their healing potential, this selective pressure would be reduced, with a consequent weakening of such capacities over generations.

For a sense of how differently shamanic people view what most contemporary psychiatrists would diagnose as severe mental illness, it's helpful to look at the life story of a Lakota shaman named Black Elk, as told to the poet John Neihardt, and published as a collection of transcriptions of their conversations in *Black Elk Speaks*, in 1932. The book has become a classic of American Indian literature.

The basic facts of Black Elk's life are astounding. He was almost an adolescent before he saw a white person, but within a few years, white people had overrun his culture. By any standards, this man experienced almost inconceivable psychological stress. He had witnessed the utter destruction of his culture, and the murders of his people's leaders and, then, in the hope of discovering "some secret of the (whites) that would help (his) people somehow," he'd joined Buffalo Bill's Wild West show and traveled to Chicago, New York, London, and Paris.

At the age of five, before the period of great drama in Black Elk's life had commenced, he began hearing voices: "I was out playing alone when I heard them. It was like someone calling me, and I thought it was my mother, but there was nobody there. This happened more than once, and always made me afraid, so that I ran home." Stephen Larsen, a psychotherapist and authority on world mythology, distinguishes "mythologized" cultures like the Lakota's from "demythologized" cultures like our own. In the former, "Mythic meaning and social meaning are . . . brought together rather than separated, and the archaic type of thinking is fused with mythic images and social realities." By contrast, Larsen believes that the civilized mind neglects mythological ways of understanding the world, so these images and insights are suppressed, emerging only in fantasies and dreams.

The "great vision" that Black Elk remembered throughout his life was preceded by severe physical symptoms. One day, for no apparent reason, both his legs began to hurt. By the next morning, the boy was unable to walk at all, and his arms, legs, and face were all swollen. While suffering from this condition, he had an extended vision that included conversations with some of his people's gods, and being granted the power to heal others, to communicate with animals, and even to travel outside his body.

A classically trained Western mental health worker would diagnose anyone with experiences similar to these as psychotic, probably schizophrenic. It's likely that they'd be told there was no cure for the condition, prescribed powerful antipsychotic medications, and possibly institutionalized for the rest of their lives. But that's not what happened to Black Elk. Instead, his concerned parents called on a traditional Lakota shaman named Black Road, who sat alone with the boy in a tepee and asked to hear about his vision. "I was so afraid of being afraid of everything that I told him about my vision, and when I was through he looked long at me and said: 'Ah-h-h-h!,' meaning that he was much surprised." Black Road told the boy that now he knew what the trouble was: Black Elk must respect the voices and "perform this vision for your people upon earth. . . . Then the fear will leave you; but if you do not do this, something very bad will happen to you."

Everyone who knew this troubled boy agreed to participate in enacting the images and sounds that had been tormenting him— down to the smallest details. They set up a sacred tepee and spent all day painting the hides with images from Black Elk's vision. They stayed up all night learning the sacred songs the young man had heard in his vision. Sixteen young men rode horses of particular colors, four each from each of the four sacred directions; four young girls in the village played their part in the enactment, as well as six old men. People painted their faces and bodies according to the boy's specifications. They gathered food and played drums in unison.

This troubled young man, haunted by hallucinated voices and disturbing visions, was embraced by his community in an intimate, supportive, and loving way. They came together to bring to life, in as much detail as possible, the images and sounds that had been tormenting him for years.

This is not noble savagery. There is plenty of self-interest involved here. In shamanic societies, it's understood that a person capable of moving between worlds can be a great asset—a healer who will spend the rest of his or her life using this capacity to help others. As explained by psychiatrist Roger Walsh, while "Western psychiatry has a long history of viewing mystics as madmen, saints as psychotics, and sages as schizophrenics," in traditional societies, these experiences may be seen as "proof that (one) is destined to be a shaman." Such a young person "is understood by the tribe to be undergoing a difficult but potentially valuable developmental process. If handled appropriately this process is expected to resolve in ways that will benefit the whole tribe and provide them with new access to spiritual realms and powers."

The sort of cathartic healing Walsh described is what happened in Black Elk's case. At the climax of the enactment of his vision, the boy looked up at the sky and, as he recounted many years later, "As I sat there looking at the cloud, I saw my vision yonder once again. . . . I looked about me and could see that what we then were doing was like a shadow cast upon the earth from yonder vision in the heavens, so bright it was and clear. I knew the real was yonder and the darkened dream of it was here." This experience transformed the terrified boy into a man able to withstand psychological pressures beyond imagining.

Not all mental health crises can be resolved by the kind of interventions shamanic societies can offer. Some conditions are organic, due to genetics, complications in pregnancy, head trauma, and so on. In such situations, modern psychiatric interventions can be lifesaving. But most of the suffering we see around us today is due to social causes that can and must be addressed before they manifest as mental illness: economic insecurity, misinformed parenting practices, oppressive educational systems, war and domestic violence, shame concerning sexuality and our bodies, absurd notions of beauty and success calculated to keep us always dissatisfied with ourselves and our lives. No pills will ever address these sources of our distress.

Our civilized impulse is to remove or weaken the perceived danger: kill it before it kills us. We place babies in sterile incubators; send our children to schools with armed guards, metal detectors, and teachers who are legally forbidden to touch even a crying child; drop bombs across the globe that create more potential terrorists than they kill; and administer drugs that quiet voices we should be listening to. It hasn't worked, and never will—a fact to which we seem to be slowly waking up. Our survival depends not on eliminating the dangers of life but on relearning to embrace and acknowledge that which terrifies us—including altered states of consciousness.

– Turn On, Tune In, Get Better –

What if I told you that scientists had discovered a new kind of drug that is nontoxic, nonaddictive, inexpensive, and far out-performs anything doctors currently use to treat severe anxiety, addictions to alcohol, tobacco, cocaine, and opiates, and the physical and psychological suffering of people dying from cancer? In fact, they are very old drugs and are basically free. They're called psychedelics.

In nearly every society, psychedelics have been considered among the greatest gifts bestowed on humanity by the gods. From ayahuasca in the Amazon to the peyote used by Huichol Indians in Mexico to iboga in Africa to amanita mushrooms in Siberia and India to LSD in the offices of European and American psychiatrists in the 1950s, psychedelics have been considered sacred substances to be used with reverence, ritual, and respect. The one glaring exception is here and now, where possession of these nonaddictive, nontoxic substances can result in spending the rest of your life in a cage. Hyperbole? I wish it were.

Due to minimum mandatory sentencing guidelines imple-mented as part of the Reagan administration's "tough on drugs" policies in the 1980s, people convicted of distribution of these drugs have been routinely imprisoned for far longer than convicted murderers. Timothy Tyler, for example, was sentenced to a double life term in 1992 for selling LSD to a friend. He was twenty-three years old. The following year, Bob Riley received a life sentence for selling psilocybin mushrooms at a Grateful Dead concert. The judge who imposed the sentence on Riley, U.S. District Court judge Robert Longstaff, a Reagan appointee, regretted what he was forced to do: "The mandatory life sentence as applied to you is not just, it's an unfair sentence," he said at the sentencing hearing, before sending Riley away in chains. These examples are representative of a society that routinely imposes life sentences on people who distribute the same substances that virtually every other society has celebrated and cherished.

But that may be changing. When I first met Rick Doblin, in the mid-1990s, I thought he was an admirable dreamer. In the early 1980s, Rick was a young therapist-in-training who'd had some experience with MDMA, which was still legal at the time. MDMA (later known as Adam, ecstasy, and Molly, among other names) was being used by an informal network of therapists, mainly on the West Coast of the United States. A chemist at Merck first synthesized the drug in the 1920s but shelved it after scientists failed to notice any clinically significant effects in animal studies. When the drug was rediscovered by the legendary chemist Sasha Shulgin in the late 1970s, many therapists referred to it as an "empathogen," due to the empathy and compassion most people feel when they take it. This quality made MDMA popular among therapists, particularly those working with patients struggling with debilitating anxiety and couples whose accumulated rage and hostility impeded productive communication.

When MDMA became popular in dance clubs in the early 1980s, people started showing up in emergency rooms complaining of overdose, and it became clear that the federal government was going to block the recreational use of MDMA. But because deaths were rare, neurotoxicity hadn't been demonstrated, and therapeutic benefits were well established, the DEA held a series of hearings to determine whether the drug had any legitimate therapeutic use. The judge recommended that MDMA be classified as a Schedule III substance, due to its established history of successful medical use, but the administrator of the DEA overruled the recommendation and classified the drug as Schedule I (no medical use). When

Harvard psychiatrist Lester Grinspoon sued the DEA to try to force it to recognize the well-established medical uses, a federal court agreed, vacating MDMA's Schedule I status. But just a few weeks later, the DEA overruled the federal court and returned MDMA to Schedule I. The federal government was determined to claim MDMA had no potentially beneficial properties.

At this point, Rick Doblin began what became his life's work: finding a way to responsibly and legally reintroduce MDMA and other drugs with psychedelic properties into mainstream American clinical practice. He founded the Multidisciplinary Association for Psychedelic Studies (MAPS) in 1986, and enrolled at Harvard's Kennedy School of Government, where he wrote his dissertation on the regulation of the medical uses of psychedelics and marijuana.

What Doblin proposed to do was radical, but not unprecedented. Psychedelics were widely used all over the world, until the practice was stamped out through vicious suppression by religious authorities who felt that freely available plant-based contact with the divine threatened their monopoly on the sacred. During the Spanish Conquest of Mexico, possession of psilocybin mushrooms, which the Aztecs called *teonanácatl* ("flesh of the gods"), was punishable by death. The Spaniards worshipped a "jealous God," indeed.

Similarly, before the arrival of Christianity and its prohibitions, the indigenous healers of Europe sometimes used potions of *Amanita muscaria* mushrooms or toads, some species of which have two powerful drugs—5-MeO-DMT and bufotenin—in their skin. Because both the mushrooms and the toads are highly toxic, these potions weren't swallowed, but guided into the bloodstream via the mucus membranes. Many of these early healers were women, and historians have documented that one method of ingesting their "magic potions" was by dipping a phallus-shaped wand in the potion and then rubbing it against or inside the vaginal mucosa. The Christian campaign to wipe out these practices featured the literal demonization of these women, and "witches" still ride their phallic broomsticks to this day.

In the 1950s and early 1960s, many psychologists and psychiatrists believed LSD to be a "psychotomimetic" that provoked a temporary state of psychosis. These modern-day shamans took high doses of the drug so that they could better relate to their patients by having spent eight to twelve hours experiencing the same kinds of disorientation, hallucinations, paranoia, and transcendental states often reported by their patients. Researchers in Europe, the United States, and Canada conducted thousands of experiments using LSD and psilocybin to treat a wide variety of conditions, including alcoholism, obsessive-compulsive disorder, depression, and schizophrenia. Additional research was being conducted by the CIA and military scientists interested in using these substances for interrogations, to supercharge intellectual capacities, and so on. In this period, around $4 million in federal money financed more than a hundred studies of LSD in the United States. Some psychiatrists, most famously Oscar Janiger, conducted private sessions with "patients" who wanted to experience the otherworldly effects of these substances, including Anaïs Nin, Aldous Huxley, Cary Grant, Rita Moreno, and Jack Nicholson. No one, among the millions who've taken LSD, has died of an overdose. While people have, in rare cases, made ill-advised decisions that have led to their deaths, such incidents are due to ignorance about the proper use of the substance, not the substance itself.

As the country began to turn against the war raging in Vietnam, and tension grew over the violence and urban decay in America's inner cities, the Nixon administration concocted another war designed to target and neutralize the two loudest voices of protest:

hippies and black people. The "war on drugs," it turns out, was never really about the drugs. It was simply a way to marginalize and silence protest against the waste of life and treasure in Southeast Asia. John Ehrlichman, Nixon's domestic policy advisor in 1968, explained the plan to journalist Dan Baum in 1994:

> The Nixon campaign in 1968, and the Nixon White House after that, had two enemies: the antiwar left and black people. You understand what I'm saying? We knew we couldn't make it illegal to be either against the war or black, but by getting the public to associate the hippies with marijuana and blacks with heroin, and then criminalizing both heavily, we could disrupt those communities. We could arrest their leaders, raid their homes, break up their meetings, and vilify them night after night on the evening news. Did we know we were lying about the drugs? Of course we did.

In his rush to attack hippies and black people, one of America's most despised presidents torpedoed decades of accumulated research into the healing potential of psychedelics by declaring all such drugs illegal. But the splash made by psychedelics continues to ripple through the culture of the United States and the world. Without the influence of psychedelics, it's hard to imagine the quantum advances in music, art, film, and science that marked the last few decades of the twentieth century. Francis Crick, the discoverer of the DNA double helix, was a day-tripper, the Beatles went from playing "I Wanna Hold Your Hand" in monkey suits to "Strawberry Fields Forever," and Steve Jobs recalled his experiences with LSD as "one of the two or three most important things I ever did in my life."

The association of psychedelics with cutting-edge creativity continues. Many have argued it's no coincidence that the genius of the San Francisco Bay Area is a child of the marriage of technology and acid-generated free thinking consummated in the 1970s. Tim Ferriss, a well-known Silicon Valley investor and author, has said he knows lots of successful entrepreneurs who use psychedelics regularly, if not religiously. In an interview with CNN Money, Ferriss said, "The billionaires I know, almost without exception, use hallucinogens on a regular basis. [They're] trying to be very disruptive and look at the problems in the world . . . and ask completely new questions." There's good reason to adopt this approach, aside from the anecdotal evidence. According to Dr. Robin Carhart-Harris, who has conducted research using fMRI machines to study how the brain behaves under the influence of psychedelics, they "dismantle 'well-worn' networks," allowing novel communication patterns to occur.

Claims that these substances are toxic or otherwise dangerous are widespread, but without basis. In one famous example, in 1967, *Science* reported that LSD damaged chromosomes, despite the fact that the research being reported *was based upon a single subject*. This unscientific claptrap was immediately picked up by national media. In articles with alarming headlines such as "The Hidden Evils of LSD," journalists declared, "New research finds [LSD] is causing genetic damage that poses a threat of havoc now and appalling abnormalities for generations yet unborn." For the next five years, both scientific and popular literature amplified and circulated this hysterical nonsense, drawing a terrifying link between LSD and future birth defects. All this, *based upon a single study of one person.*

Four years later, in 1971, *Science* published a follow-up study admitting, "Pure LSD ingested in moderate doses does not damage chromosomes in vivo, does not cause detectable genetic damage, and is not a teratogen or a carcinogen in man." But no headlines trumpeted this news. Even now, almost half a century later, many people remain convinced that there is a well-demonstrated link between the use of LSD and chromosomal damage.

In addition to being nontoxic to bodily organs, psychedelics appear to pose little danger to mental health. While such substances should always be approached with caution and respect, an assessment of psychedelic drugs conducted by the World Health Organization failed to cite even a single example of any harm from naturally occurring psychedelics (and only a handful of anecdotes related to LSD). A large-scale study published in the *Journal of Psychopharmacology* in 2015 looked at 130,000 American adults and failed to find any evidence linking the use of psychedelic substances with mental health problems of any kind. The researchers "found no significant associations between lifetime use of psychedelics and increased likelihood of past year serious psychological distress, mental health treatment, suicidal thoughts, suicidal plans and suicide attempt, depression and anxiety." They conclude the study by observing that "it is difficult to see how prohibition of psychedelics can be justified as a public health measure."

Psychedelics do carry risk, though. Because of the profound perceptual alterations they can trigger, psychedelic experiences can be frightening and disorienting. People already suffering from serious mental health conditions, a shaky sense of reality, or damaged self-worth may find these disruptions very distressing. Similarly, psychedelics' tendency to pull away the veils of self-delusion can create difficulties for anyone going through difficult times, as it becomes increasingly difficult to ignore the reality of one's life. Yet when the United Kingdom's Independent Scientific Committee on Drugs assessed the relative dangers posed by various substances to the user and to others, alcohol topped the ranking, while LSD and mushrooms came in last. In the words of Professor David Nutt, who led the committee, "It's virtually impossible to die from an overdose of them; they cause no physical harm; and if anything they are anti-addictive, as they cause a sudden tolerance, which means that if you immediately take another dose it will probably have very little effect, so there is no incentive to take more."

Roland Griffiths, a psychopharmacologist at Johns Hopkins University School of Medicine, is mystified by our culture's panic around psychedelics: "We ended up demonizing these compounds, [but] can you think of another area of science regarded as so dangerous and taboo that all research gets shut down for decades? It's unprecedented in modern science." In Griffiths's study, thirty-six volunteers took a pill containing either psilocybin or a stimulant that would provoke some mild physiological effects. "When administered under supportive conditions," the researchers concluded, "psilocybin occasioned experiences similar to spontaneously occurring mystical experiences." Participants felt that these were among the most affecting experiences in their lives, comparable to the birth of a child or the death of a parent. Two-thirds ranked the psilocybin session as being among the top five most spiritually significant experiences of their lives and another third said it was the most significant spiritual experience they'd ever had.

Katherine MacLean, a psychologist in Griffiths's lab, conducted a follow-up study, finding that the subjects enjoyed positive and lasting changes in their personality—despite the conventional view that personality is more or less set by the age of thirty. More than a year after their experience with psilocybin, many of the subjects still showed significantly greater tolerance, flexibility, and creativity. "I don't want to use the word 'mind-blowing,'" Griffiths said, "but, as a scientific phenomenon, if you can create conditions in which seventy per cent of people will say they have had one of the five most meaningful experiences of their lives? To a scientist, that's just incredible."

Griffiths and his colleagues have also been studying the potential of psilocybin to treat tobacco addiction. Because these are still early trials, the sample sizes are small, but the results have been explosive: Six months after receiving treatment, 80 percent of the subjects (all of whom had tried to quit smoking multiple times before) were still not smoking. Compare this to a success rate of just 7 percent for nicotine-replacement therapy, currently the most successful approach.

Two other drugs long familiar to shamanic societies that are beginning to receive a lot of attention—particularly for their potential in helping people break out of addictive behavior patterns—are ayahuasca and iboga. Ayahuasca (also known as *yage*, and a word whose etymological roots in the Quechua language refer to "the vine of the soul") is a concoction made of two plants native to the Amazon basin: *Banisteriopsis caapi*, which is a vine, and the leaves of the *Psychotria viridis* plant. Ayahuasca has a range of powerful healing actions. The principal psychoactive ingredient is dimethyltryptamine (DMT), a substance naturally produced in the human brain. It has been detected in our blood and cerebrospinal fluid as well. While DMT has mostly been studied with a focus on its psychedelic effects, the substance is also involved in our dreams, near-death experiences, and psychotic episodes. Recent research suggests that the healing potential of DMT may be considerable. A recent review paper called "The Therapeutic Potentials of Ayahuasca: Possible Effects against Various Diseases of Civilization" was published in the journal *Frontiers in Pharmacology* (March 2016). The authors of this paper outline some of the demonstrated benefits people have derived from their experiences with the brew and conclude that it is best understood "from a bio-psycho-socio-spiritual model" and that ayahuasca "may act against chronic low grade inflammation and oxidative stress." The authors do their best to maintain an understated, scientific tone while pointing to a potentially massive medical breakthrough: "Altogether, no other receptor has ever been associated with so many different diseases as the Sig-1R [which bonds to DMT]. It has so far been implicated in illnesses like Alzheimer's disease, Parkinson's disease, cancer, cardiomyopathy, retinal dysfunction, perinatal and traumatic brain injury, frontal motor neuron degeneration, amyotrophic lateral sclerosis, HIV-related dementia, major depression, and psychostimulant addiction." A deeper understanding and appreciation of ayahuasca seems essential to reconfiguring the human zoo for a healthier, happier society.

The word "iboga" apparently comes from the Tsogo verb *boghaga*, "to care for." Like ayahuasca, it has a long history of having been used in shamanic ceremonies, most famously in coming-of-age rituals among practitioners of the Bwiti religion in the West African nation of Gabon. Today, it's becoming well known as a difficult but highly effective aid to breaking out of addictive behavior patterns—particularly those involving opiates. Like ayahuasca, iboga isn't a party drug, as the hallucinogenic effects are far too intense to be "fun" and last anywhere from twenty-four to forty-eight hours. Furthermore, iboga can be dangerous. A handful of deaths have occurred at clinics, but autopsies have found that, in all cases, the deceased either suffered from pre-existing, undiagnosed cardiac issues or had taken other substances—generally cocaine—together with the iboga.

While the specific mechanisms of action aren't well understood (largely because of the difficulties of conducting research on an illegal substance), an iboga experience leaves many addicts feeling as if their long-standing cravings have been wiped clean—as if their brains have somehow been reset. One man who'd been addicted to heroin for several years before spending the last of his student loan money on a session at a clinic in Tijuana, Mexico, told me that after a single experience with iboga he felt as if he'd never taken heroin in his life. "There was no gravitational pull at all," he said. "If I relapsed, it would be like starting again from zero, not like going back to where I wanted to be."

According to the most current reports from the CDC, overdose deaths involving prescription opioids quadrupled between 1999 and 2014, as did sales of these drugs. And the numbers continue to increase. In 2017, more Americans died from drug overdose than died from any cause in the entire Vietnam War (sixty-four thousand versus fifty-eight thousand). The vast majority of those overdoses are from opioids—pain medications we overprescribe because we insist on treating superficial symptoms rather than underlying structural problems in how we live our lives. And while this epidemic continues, the federal government continues to prohibit the clinical use of ayahuasca and iboga—the most effective addiction treatments known.

Researchers have found that psilocybin is so effective in alleviating the existential fears of the dying that just a single dose produces immediate and dramatic reductions in anxiety and depression in people suffering from terminal cancer. The psychological benefits were undiminished even six months later in patients who survived that long. One of the scientists involved in this research was so amazed by the results that he questioned whether they could be real. "I thought the first ten or twenty people . . . must be faking it. . . . People who had been palpably scared of death—they lost their fear. The fact that a drug given once can have such an effect for so long is an unprecedented finding. We have never had anything like it in the psychiatric field." The dumbfounded researcher is Dr. Stephen Ross, associate professor of psychiatry and child and adolescent psychiatry at the New York University (NYU) School of Medicine.

Another scientist involved in this study, Dr. Anthony Bossis, is a specialist in palliative care research. Bossis believes that psilocybin-assisted therapy could help patients by triggering a cathartic state of enhanced spiritual awareness. "Psilocybin and other psychoactive organic compounds have been used for millennia and have reliably been shown to activate what is known as the mystical experience in humans," he said. "The mystical experience has been shown to improve a patient's existential well-being and [his] ability to reframe the impact cancer has on [his] life by giving [the patient] an increased appreciation of time living. The patient recognizes that [he is] not dying per se; [he is] living, until the moment of death. Ultimately, the patient fears death less and embraces life more, becoming an active participant in life and enriching [his] interpersonal relationships, which is one of the first casualties in advanced cancer."

While the diminution of existential fear may well be "unprecedented" in modern medical research, it's well established in human experience. Archaeological evidence for the ritualistic use of psychoactive mushrooms dates back at least fifty-seven hundred years—predating, by thousands of years, all the major organized religions in the world today. Historians have suggested that mushrooms may well have been the "soma" mentioned in the ancient Hindu Vedas and "nepenthe," the "drug of forgetfulness," mentioned by Homer in *The Odyssey*. And the long-standing use of such substances in shamanic societies has been documented from the Arctic Circle to the steamiest jungles of the Amazon.

David Nutt, a celebrated British psychiatrist and neuropsychopharmacologist as well as a distinguished researcher, had long been an advisor on drug policy to the British government when he published the results of a study of comparative drug harms in *The Lancet* in 2007. He was quickly fired. As in the United States, ignorant political hacks in the United Kingdom insist on demonizing relatively harmless substances that produce feelings of empathy, love, and compassion, while they pickle their livers in single-malt Scotch. Scientists who dare to disagree publicly—even those at the pinnacle of their fields—risk professional suicide. Drug policy is still more politics than science, but thanks to courageous scientists and academics such as David Nutt, Andrew Weil, Charles Grob, Stanley Krippner, Rick Doblin, and many others who have openly discussed the powerful healing potential of these substances and refused to be intimidated into silence, prohibitions against research and clinical use are beginning to loosen.

Change is coming, but it's achingly slow. Thousands of what can only be considered political prisoners still languish in prisons around the world—convicted only of facilitating the use of substances our species has used to alleviate suffering and enrich consciousness for millennia. Fear of the healing and educational potential of psychedelics has begun to fade, partly because highly respected mainstream figures such as neurologist Oliver Sacks, Steve Jobs, and Nobel laureates Francis Crick and Kary Mullis have openly discussed the lasting value of their experiences with LSD, psilocybin, and other psychedelics. In 2015, at the largest professional gathering of psychiatrists in the world, the president of the American Psychiatric Association, Dr. Paul Summergrad, openly credited an early LSD trip with helping him decide to devote his life to the field of psychiatry.

*   *   *

Foragers see the world as spiritually alive, welcoming, and generous. Farmers tend to see it as inanimate, forbidding, and reluctant. The gods of foragers are multiple, benevolent, and directly accessible by anyone; the God of farmers is solitary, angry, and jealous. Whatever minimal property foragers possess is to be shared without reservation. Agriculturalists are taught to hoard property and defend it to the death. While foragers tend to see one another as companions in mutually beneficial relationships, farmers tend to view one another as rivals in a zero-sum situation. There are many ways to illuminate this difference between our timeless essence and our current predicament. In his cult classic *Ishmael*, for example, Daniel Quinn distinguished "the leavers" from "the takers." At the risk of sounding hopelessly Rousseauian, I'd suggest an equally clarifying polarity is love versus fear.

In her classic book *On Death and Dying*, Elisabeth Kübler-Ross identified five stages of grief that most people seem to pass through when processing loss—whether it be the loss of a relationship, close friend, job, or life itself. Once you learn the stages, you'll start seeing them everywhere.

Denial ("The lab must have made a mistake.")

Anger ("Why me? This isn't fair!")

Bargaining ("I promise I'll change.")

Depression ("What's the point? I'm so tired.")

Acceptance ("I can handle this.")

DABDA. The acceptance stage is attained when fear is vanquished and love once again becomes a possibility. The earlier stages are all expressions of a progressing panic, focused on what's being lost rather than what remains, or is being gained. Learning to accept the inevitability of what we fear most is the essential step on the path to a life worth living. Millennia of struggling against this uniquely human knowledge have transformed a relatively relaxed, egalitarian primate into a beast that is often aggressive, frustrated, and fearful: We've gone from grasshopper to locust. "In the end fear casts out even a man's humanity," wrote Aldous Huxley, in *Ape and Essence*. "And fear, my good friends, fear is the very basis and foundation of modern life."

The Narrative of Perpetual Progress is ubiquitous because it serves the purposes of a modern world built on fear. We learn to work toward everlasting life by praying to the right god, purchasing the right stuff, going to the right schools, taking the right supplements, doing the right exercises, and fighting for the right army. At the same time, we're reminded that it's a cruel world out there, and that we're all helpless. We rush onward, trampling what's left of the Garden, fleeing inchoate specters of hunger, abandonment, terrorism, economic collapse, police and criminals, nuclear meltdown, volcanic upheaval, asteroids, and death. Always death.

The mysterious and much-needed power of psychedelics to help cast off the fear of dying may help us move toward a mature acceptance of what life actually offers and requires of us. This insight, so essential to living a life of authenticity and integrity, threatens the false narrative of civilization so deeply that for centuries indigenous healers who used such substances have been condemned by the civilized as witches or heretics and burned alive. Even today, we condemn harmless teenagers to decades in cages because they brought magic mushrooms to a muddy concert.

The mantra of the angry revolution of the 1960s was "turn on, tune in, drop out." Hippies long since grown up and now growing old know that the revolution isn't about tearing down the world so much as protecting it and giving future generations even a long shot at survival. We need to take our wisdom where we find it, and that clearly includes psychedelics. "It's not 'turn on, tune in and drop out'" anymore, muses Doblin. "It's 'turn on, tune in and take over.'"

But how "real" is this psychedelic worldview shared by foragers and freaks?

– On Holy Ghosts –

When asked to define "reality," the famous science-fiction writer Philip K. Dick said, "Reality is that which, when you stop believing in it, doesn't go away." It's a great line, but it misses an important aspect of reality: the part that is sustained by belief, but that is no less real for that. Confronted by ghosts knocking over candles, better to put out the fire before arguing about the reality of ghosts. When it comes to mystical experiences, the wisest course is to judge their results rather than to be derailed by our current inability to explain their mechanism of action. In an interview published in *Science*, David Nichols, one of the founders of the Heffter Research Institute, a major supporter of psychedelic research, was asked about the "reality" of therapeutic approaches that incorporate psychedelics. He said: "If it gives them peace, if it helps people to die peacefully with their friends and their family at their side, I don't care if it's real or an illusion." Indeed, and the distinction between "real" and "illusion" begins to dissipate in such a setting.

The so-called scientific worldview is often limited by an unwillingness to accept the reality of the inexplicable. To dismiss the tangible, measurable, predictable, life-altering effects of psychedelics as some sort of "hippie nonsense" is to shoot ourselves in the foot because we don't understand how bullets work.

In 1977, the great theoretical physicist David Bohm described his sense of the circular swirl formed by our beliefs and what we experience as reality: "Reality is what we take to be true. What we take to be true is what we believe. What we believe is based upon our perceptions. What we perceive depends on what we look for. What we look for depends on what we think. What we think depends on what we perceive. What we perceive determines what we believe. What we believe determines what we take to be true. What we take to be true is our reality." (That's worth reading a few times.)

In *The Geography of Madness: Penis Thieves, Voodoo Death, and the Search for the Meaning of the World's Strangest Symptoms,* Frank Bures surveys some of the bizarre yet demonstrably "real" ailments that people suffer from precisely because they believe in them:

> There was amok, from Malaysia, in which a person brooded for a period of time, then went on a random homicidal rampage, but had no memory of the events afterward. . . . In Japan, certain people suffered from taijin kyofusho, a terrifying fear of other people's embarrassment (not their own). . . . In Cambodia, people suffered from khyâl cap, or "wind attacks," in which khyâl, a "wind-like substance" believed to flow alongside blood, rushes to the head and causes all kinds of problems, including dizziness, shortness of breath, numbness, fever, and so on. . . . Indian men were at risk of dhat syndrome, in which they lost weight and felt fatigue, weakness, and impotence due to loss of semen—one of seven essential bodily fluids in Ayurvedic medicine. In some parts of the country, they also contracted koro: In 1982, eighty-three men and women in Lower Assam rushed to hospitals with a "tingling" in their lower abdomen and a fear that their testicles or breasts were shrinking.

In addition to these imaginary-yet-real ailments, human beings are notoriously adept at finding religious significance in the most unlikely places. No journalist seems capable of *not* referring to the

Higgs boson as the "God particle" even though it has nothing to do with God. The term was coined when a physicist—Nobel laureate Leon Lederman—proposed naming his book on the Higgs *The Goddamn Particle*, because it had been so difficult to find. His editor made a slight adjustment to the title, and now the world is stuck with it. What a difference a *damn* makes.

In an essay called "The Myth of Mechanism," published in 2001, T. V. Rajan explained his frustration with medical research that refuses to acknowledge the inexplicable. Rajan, who is professor of immunology and experimental pathology at the UConn Health Center, writes that there are two different aspects to research: to determine whether a phenomenon exists and whether it "operates by a mechanism that can be comprehended in the context of our current knowledge of human physiology and behavior." Rajan sees too many of his colleagues refusing to acknowledge the existence of what they cannot explain—a mistake, he thinks, because the question of whether we understand *how* something exists should not be confused with the question of *whether* it exists.

Rajan gives several examples of important medical advances where the mechanism of action was (and, in some cases, still is) totally unexplained: digitalis (used to treat various heart conditions), diethylcarbamazine (used to treat lymphatic disorders), chloroquin (antimalarial)—all of which are "drug[s] in search of a mechanism," in Rajan's words. He could also have mentioned the creation, in 1796, of the first vaccine, by Edward Jenner, who had no idea how the cowpox to which he'd exposed an unwitting child had conferred immunity to smallpox. It was just a hunch. Another hunch led William Coley to inject cancer patients with a *Streptococcus* bacterium a hundred years later, with surprisingly (and inexplicably) positive results. Testifying before a Senate committee in 1971, pathologist Sidney Farber explained, "The history of medicine is replete with examples of cures obtained years, decades, and even centuries before the mechanism of action was understood for these cures."

Rajan calls it "hubris" to dismiss phenomena because we cannot explain how they happen. "We seem to understand so little about human biology and physiology," he writes, "that a vast majority of what happens to us can simply not be understood, at least with today's knowledge base. . . . If nothing else, humility dictates that we appreciate that there are more things in heaven and earth than are dreamt of in our philosophy."

Science is certainly one of the strongest lights ever to illuminate the known universe. But the light of science can be shadowy and spectral. Those who insist that nothing exists beyond that which is scientifically demonstrable are like children who cover their eyes and imagine the world disappears because they cannot see it.

The worm at the core of the recent catechism being preached with such fervor by so-called New Atheists is the conviction that because various elements in someone's religious beliefs are demonstrably untrue (for example, that all the animals in the world are descendants of Noah's ark, that Earth is only seven thousand years old, or that this God is better than that God), their religious experience is, by extension, unreal. The misconception underlying this view presumes a digital universe. Something is either true or it isn't. Yes or no. On or off. Dead or alive. Zero or one.

But every life is full of moments when objective reality and experience mitigated by belief swirl together as inseparably and deliciously as coffee and cream. Placebos aren't "real" but they are effective in relieving pain and depression—particularly in American patients. We don't know how a placebo works, but there's no doubt that it does—often as well as or better than the most expensive pharmaceuticals. And belief is at the core of the effectiveness of the placebo response. When you stop believing in this reality, Mr. Dick, it disappears.

A comprehensive review of more than two hundred published studies found that patients' religiosity was associated with better health outcomes. Jeffrey Levin, the author, writes that "results pointed to a mostly salutary or protective epidemiological effect of religiosity, regardless of the religious measure used or the outcome under study, and this relationship manifested in study populations regardless of age, sex, race, ethnicity, nationality, study design or the period of time during which the study was conducted."

Religious faith may be inspired by fanciful stories, but it can make the intolerable tolerable. Love may add up to nothing more than a sustained bout of mutual idealized projection, but it's what we live for. Little mushrooms growing in clumps of cow shit can provoke mystical experiences so otherworldly that we're suddenly able to get up and walk away from decades of self-destructive behavior or settle into our final journey with calm acceptance. You can't get more real than that.

In addition to reconfiguring our relationships with physical and ego death, people are finding many ways to bring their modern lives into alignment with ancient, eternal human appetites and trajectories.

– Past Progressive –

So we beat on, boats against the current, borne back ceaselessly into the past.

—F. Scott Fitzgerald, *The Great Gatsby*

When you're lost, a step back may be a step in the right direction. Every day, more people conclude that the approach to life promoted by the central myths of civilization are generating loneliness, confusion, anxiety, and despair for many of us. Practically every aspect of modern life is up for re-examination, and, across the board, we're looking to the original environment of our species for guidance: natural childbirth, free-range, cruelty-free meat, organic fruits and vegetables, horizontal business organization, the sharing economy, nonbinary sexualities and flexible relationship configurations, LGBTQ rights, minimalist shelter and personal finance, complementary medicine, psychedelic-assisted psycho-therapy . . . every one of these growing trends and many more like them are rooted in paleo principles. We've already looked at some of the ways an understanding of undomesticated human life informs birth, parenting, work and our relationship to money, psychedelic psychotherapy, and how we approach death. Dozens of books and documentary films are coming out every year exploring how these same principles are changing how people look at what kind of home they want to live in, how to maintain or regain their health, and how to manage their finances.

*   *   *

Steven Johnson calls himself a "peer progressive." "We believe in social progress," he wrote, in *Future Perfect: The Case for Progress in a Networked Age*, "and we believe the most powerful tool to advance the cause of progress is the peer network." Johnson believes that "the key to continued progress lies in building peer networks in as many regions of modern life as possible: in education, health care, city neighborhoods, private corporations, and government agencies," but he harbors no illusions about where we are right now: "Twenty-first-century marketplaces are dominated by immense, hierarchically organized global corporations—the very antithesis of peer networks." Still, those immense corporations are far more vulnerable than they seem, and many are already on their way out.

Airbnb and similar networks are diverting market share from hotels all over the world, while taxis are being displaced by ride-sharing apps such as Lyft—which will soon have to compete with self-driving cars already being tested on the roads of Silicon Valley.

One of Johnson's favorite examples of the new, decentralized economy is Kickstarter, where "both the ideas and the funding come from the edges of the network; the service itself just supplies the software that makes those connections possible. . . . There are no experts, no leaders, no bureaucrats—only peers." While the company is itself a classic example of modern capitalism, what it has created is an alternative to capitalism that harkens back to the ancestral form of human exchange: a gift economy. How powerful is this approach? According to the company's public information, in less than a decade since the site launched in 2009, $3.5 billion has been pledged by over 14 million people to support creative projects.

What Johnson calls "peer networks" are essentially scaled-up modern reflections of the social networks in which our ancestors lived for hundreds of thousands of years. "When a need arises in society that goes unmet," writes Johnson, "our first impulse should be to build a peer network to solve that problem."

Once you start thinking along these lines, you'll begin to see this struggle between forager values and civilizational values all over the place. A so-called progressive agenda often aligns with forager values: a more equitable distribution of resources, assistance for the vulnerable, respect and autonomy for women (including equal pay and reproductive rights), increased funding for health care and educational programs, acceptance of all religions, and so on. A more conservative agenda often aligns with such agricultural values as individual rights superseding those of the community, paternalistic male control over women's sexual behavior, expansionist militarism, exaltation of wealth, and monotheism.

It may seem far-fetched to link a very contemporary company like Kickstarter with hunter-gatherers, but Johnson would agree that there's a direct connection: "There is something in the collaborative, egalitarian structure of these systems that resonates with the human mind, an echo of our deep history as a species. . . . The social architectures of the Paleolithic era—the human mind's formative years—were much closer to peer networks than they were to states or corporations."

This kind of egalitarian, horizontal network is made possible by the internet and its associated gadgetry. The potential implications are exciting and include voting and campaign contributions via smartphone, more dispersed independent publishing and journalism, crypto-banking and currency exchange, rapid response disaster relief organizations, remote medicine, and inexpensive education.

To be candid, when I first picked up Johnson's book, it was with the intention of exposing some of the ways his optimism about the future is mistaken or poorly argued (see the previous discussion of *The Rational Optimist*). I mean, *Future Perfect*? Please. But after reading his upbeat though not at all naïve arguments, I was forced to admit that he might be onto something. Nobody can foresee the changes that will come as the human swarm reconfigures itself with interconnectivity at a level unimaginable just a few decades ago. As futurist Kevin Kelly puts it, "Running a system is the quickest, shortest, and only sure method to discern emergent structures latent in it. There are no shortcuts to actually 'expressing' a convoluted, nonlinear equation to discover what it does. Too much of its behavior is packed away. . . . The most unexpected things will brew in this bionic hivelike supermind."

It's still early in the current transformation. While the values embedded in peer-progressive organizations are innate to our species, the number of groups taking advantage of new technologies that empower these appetites is still small. Because these are the non-Hobbesian, prohuman principles that our ancestors lived by, they will consistently be more appealing and resonant than top-down, someone-else-knows-best, shut-up-and-do-what-you're-told organizational structures. Peer networks reflect the *deepest* human values, nurtured over the millions of nights our fiercely egalitarian ancestors sat around the fire together, telling stories, enjoying the presence of old friends, deciding what to do tomorrow. To the extent that these hypermodern, technology-enabled networks replicate and unleash the primordial human impulses of our ancestors toward trust, faith, and mutual compassion, we may be entering a future that's a worthy reflection of our past.

A Necessary Utopia

The world is now too dangerous for anything less than utopia.

—R. Buckminster Fuller

When journalist Bill Moyers asked Isaac Asimov about the relationship between soaring population and "the dignity of the human species," Asimov was unequivocal. "It will be completely destroyed," he said. "The same way democracy cannot survive overpopulation, human dignity cannot survive it. Convenience and decency cannot survive it. As you put more and more people into the world, the value of life not only declines, it disappears." It sometimes seems as if there is a limited quantity of quality of life in the world, and as global population continues to soar, there's less to go around. With 100 million people on the planet, there'd be plenty of fresh water, fish, space, and energy for all. But the economies in which we're currently trapped thrive on growth—even at the expense of human well-being. Endless growth is the ideology of conventional economics and the cancer cell.

Still, despite the sheer volume of grumpiness you've read so far, I'm not without hope for our species—which is not to say I'm optimistic. Hope embraces the unknown and unknowable, while optimism is a belief that everything was, is, or will be fine. I am convinced that everything wasn't, isn't, and probably won't be fine. But I like to think I may be wrong about that. On very good mornings, I sometimes think we *may* be on the verge of moving into something like a utopian age. Stranger things have happened. Of course, a hard-eyed reading of history still suggests things are going to get a lot worse before they get better. We seem to be walking a razor's edge between total economic or ecological collapse on one side, with all the usual apocalyptic flourishes, and, on the other, the continued merging of technology and human biology until we are enslaved or absorbed by our creation. But I think there's still a path that leads toward home. The future I imagine (on a good day) looks a lot like the world inhabited by our ancestors—which makes a certain kind of sense, as many journeys end with a return to where they began.

The thesis of this book is that the truest, most lasting forms of progress are often those that are built upon an understanding of the past. "Reforms by advances," Jung wrote in *Memories, Dreams, Reflections,* "that is, by new methods or gadgets, are of course impressive at first, but in the long run they are dubious and in any case dearly paid for. They by no means increase the contentment or happiness of people on the whole.... Reforms by retrogressions, on the other hand, are as a rule less expensive and in addition more lasting, for they return to the simpler, tried and tested ways of the past."

It's hardly surprising that we'd seek future guidance in our past. How our species lived in the wild tells us how best to design our modern zoo. We may be on the cusp of a future unimaginable even a few decades ago, a future in which our species slips many of the constraints that have shaped human history since Göbekli Tepe was buried in trash.

– The Upside of Armageddon –

Man is at bottom a dreadful wild animal. We know this wild animal only in the tamed state called civilization and we are

therefore shocked by occasional outbreaks of its true nature:
but if and when the bolts and bars of the legal order once fall
apart and anarchy supervenes it reveals itself for what it is.

—Arthur Schopenhauer

When civilization falls away, we catch a glimpse of human nature
in the raw. When the authoritarian structures supposedly protecting us from our dark Hobbesian nature collapse into dust
and chaos, more often than not, all heaven breaks loose. In *A
Paradise Built in Hell: The Extraordinary Communities that
Arise in Disaster*, Rebecca Solnit documents how human beings
from various cultures respond to calamity—not by looting, but
by lending a hand. After reviewing the sociological literature
and hundreds of personal accounts from disaster survivors, she
concluded that "the image of the selfish, panicky, or regressively
savage human being in times of disaster has little truth to it."
Research accumulated over decades of studying how people
behave in earthquakes, floods, and bombings shows that our
behavior is the opposite of what the NPP tells us to expect. "Disaster is sometimes a door back into paradise," says Solnit, "the
paradise at least in which we are who we hope to be, do the work
we desire, and are each our sister's and brother's keeper." While
that may sound like Hallmark-card kitsch, Solnit's conclusions
are dangerously subversive. They invert the mainstream neo-
Hobbesian narrative concerning human nature and the paternalistic institutions marketed to us as protection from each other
and from our own uncivilized impulses. "Remember," the NPP
has insisted for thousands of years, "*Homō hominī lupus est*—man
is wolf to man." But that's doubly wrong. Canids are among the
most socially sophisticated, cooperative animals, and the history
of human behavior in disaster shows that we are far from brutally

selfish creatures who turn on one another the second we think we can get away with it.

Flipping the disaster narrative 180 degrees, Solnit found that "everyday life in most places is a disaster that disruptions sometimes give us a chance to change." Got that? Up is down, black is white, and earthquakes, tsunamis, and landslides aren't the true disasters; rather, they're disruptions to the ongoing, mundane disaster that most of us call "normal life."

This radical view originated with one of the founders of disaster studies, an American sociologist named Charles E. Fritz. At the end of World War II, Fritz studied the effectiveness of the Allies' bombing campaigns on the German people. From there, he enrolled at the University of Chicago, becoming director of the Disaster Research Project in 1950. Far from being some kind of fringe thinker, Fritz is a central figure in disaster studies and his conclusions represent standard thinking among disaster sociologists.

Fritz found that natural (and man-made) disasters liberated surviving victims from an oppressive normalcy: "The traditional contrast between 'normal' and 'disaster' almost always ignores or minimizes [the] recurrent stresses of everyday life and their personal and social effects," he wrote. "It also ignores a historically consistent and continually growing body of political and social analyses that points to the failure of modern societies to fulfill an individual's basic human needs for community identity."

Fritz's description of spontaneously arising human interaction in disaster bears striking similarity to normal hunter-gatherer life, in that the "widespread sharing of danger, loss, and deprivation produces an intimate, primarily group solidarity." This sense of community brings together individual and group needs, providing "a feeling of belonging and a sense of unity rarely achieved under normal circumstances." Disasters, Fritz concluded, "may be a physical hell, but they result however temporarily in what may be regarded as a kind of social utopia."

Our primordial cravings for intimate community are thwarted and twisted by the institutions that constitute civilized life. From Rat Park to Monkey Hill to Rikers Island, social conditions can either liberate a social creature's cooperative nature or twist it into confusion, anger, and violence. Fritz points to the elements of the "social utopia" disaster survivors report: feelings of group solidarity, intimate communication, and physical and emotional support. Is there any question that these feelings are lacking in our normal lives and that we yearn for them with a desperation that warps our thought and behavior? We declare fanatical allegiance to arbitrarily chosen sports teams or to street gangs that live and die over the sacred color of their hankies. We clamor toward tribalism: anything that promises group identity, mutual protection, and even a faint echo of belonging. We are starving for what our ancestors ate every day of their lives.

If scientists who study human behavior in disasters have determined that people *don't* generally panic and turn nasty in real-world crises, why is that story line repeated again and again in the media? Disaster sociologist Kathleen Tierney, who directs the University of Colorado's Natural Hazards Center, points to "elite panic," and highlights the political function of the NPP. "Elites fear disruption of the social order, challenges to their legitimacy," she says. This elite panic is characterized by "fear of social disorder; fear of the poor, minorities and immigrants; obsession with looting and property crime; willingness to resort to deadly force; and actions taken on the basis of rumor."

The indoctrination starts early. In 2005, *Time* magazine named William Golding's *Lord of the Flies* one of the hundred best English-language novels published since 1923, and it's been required reading in many American schools since the 1960s. Even if you've never read the book, the odds are you're familiar with the story of what happens to poor Piggy at the hands of some boys gone feral on a deserted island. *Lord of the Flies* is cited as if it were anthropological evidence that children will become vicious little monsters if adults aren't around to keep them in line. Hobbes for kids.

This famous fictional account of what would happen if a group of children was left to their own devices outside the protective embrace of civilization is belied by what *did* happen when a group of boys was swept up in a storm and shipwrecked on a deserted island in 1977. They didn't break into factions, smear war paint on their faces, or kill the fat kid, as anyone who read Golding's novel would have expected. Instead, they agreed to stick together, moving about the island only in pairs to ensure nobody would get lost or suffer an accident alone. They organized a rotating system so that some of them were always awake to watch for passing ships. Fifteen months later, two boys on watch spotted a passing boat, and they were all rescued.

– The End of All Our Exploring –

We shall not cease from exploration
And the end of all our exploring
Will be to arrive where we started
And know the place for the first time.

—T. S. Eliot, *Four Quartets*

In her ruminations about the "normal" brutalities of how children are raised in civilized societies, Sarah Hrdy wondered about the future of our species: "When I hear people fretting about the future of humankind in the wake of global warming, emergent diseases and rogue viruses, crashing meteorites, and exploding suns, I find myself wondering: but even if we persist, will our species still be human?" The survival of the human species, Hrdy fears, wouldn't necessarily include the survival of our *humanity*.

As always, it's now or never. Our species seems frozen at a perpetual point of no return—every step a crossroads. Civilizations have collapsed before—in fact, they all have. But none has fallen as far as ours will when it goes. Previous collapses were regional. Ours will be planetary, with nowhere to run and hide. Plenty of rivers and lakes have been overfished or poisoned over the centuries, but now we are witnessing the destruction of entire oceanic ecosystems. The atmosphere of the planet is inflamed, and our understanding of worst-case scenarios is constantly being expanded. In 2015 the strongest hurricane ever recorded—classified as a 7 on a scale that was designed to go only up to 5—ground up the coast of Mexico.

It is a truism that we live in an age of accelerating change. But nothing can continue to accelerate forever. If we look over the horizon, ahead or behind, we see clear evidence of vast periods of stability and tranquility that dwarf our brief moment of civilizational frenzy. Archaeologists have long been confounded by the tens of thousands of years when nothing much seems to have happened to signify progress. Skeletal remains demonstrate that our ancestors were anatomically modern, with plenty of mental capacity as suggested by brains that were actually a bit larger than those of contemporary humans, *but their lives weren't changing*. Artifacts show very little advancement in the design of spear points or arrowheads, burial rites, ornamentation, and so on. Why were they stuck for so long? I'd suggest that they weren't stuck at all; they were home. If necessity is the mother of invention, why is it so hard for us to surmise that they were happy and comfortable—without any apparent need for "progress"? In our world, where the present is habitually dismissed as a staging area to a better future, and disinformation concerning the long prehistory of our species is ubiquitous, it's hard to acknowledge that our ancestors' lives weren't solitary, poor, nasty, brutish, *or* short. It's nearly impossible for us to conceive that they could have been happy to stay right where they were. But this is what the evidence suggests.

Reflecting on her years with the Yequana, Liedloff remembers being perplexed and annoyed by the "irrational" way women dealt with fetching water from a nearby stream. "The women left their firesides several times a day, carrying two or three small gourds at a time," she wrote. The way was slippery and took them about twenty minutes each time. Why not move the camp closer to the river? Why not set up a more efficient system for bringing water up from the stream? Liedloff recalls that the women would often put down their gourds, shed their clothing, and wade happily into the stream. In retrospect, it became clear that taking several trips to the stream each day was no problem at all. In terms of progress, the repeated walks to the stream made no sense. But as the Yequana were content with their lives, Liedloff finally understood that they "felt no need, no pressure from any quarter, to change their ways."

A similarly beguiling absence of progress confronts us when we look to the future. The so-called Fermi Paradox is of great concern to many of the acknowledged geniuses of our age. One afternoon shortly after helping create the world's first nuclear explosion, Enrico Fermi made some calculations on a napkin over lunch with colleagues at Los Alamos. After considering the billions of stars in our galaxy with planets orbiting in the marginal range where life could arise—many of them far older than our own sun—he asked, "Where is everybody?" Given the statistically overwhelming odds that life has emerged many, many times, and that advanced intelligence and technology appear to evolve naturally once life appears, why have we seen no evidence of anybody else? Elon Musk, Stephen Hawking, and others have expressed their concern that the silence signifies a "great filter" inherent in technological development. They believe there may be a self-destruct trigger inherent in technology that has destroyed every advanced life-form before it could send out the transmissions so glaringly absent from the sky. Either they blew themselves up, poisoned themselves, or were overtaken by ruthless artificial intelligence. Looking around at our current mess—much of which is obviously due to our inability to control the gadgetry and systems we've created—none of these dark possibilities seems particularly far-fetched.

But an extraordinary little book called *Finite and Infinite Games*, by philosopher James Carse, offers another way to understand the Fermi Paradox. Carse presents a simple, yet powerful way to think about human interaction and, by extension, the potential sustainability of human societies. Most of the games played on the field of civilization are finite and zero-sum: There are clear rules; there are winners and losers; each game has a beginning, middle, and end. But the game of life is (or should be) infinite: Rules are made by players who are free to change them at any time; there are no winners and losers, just players; and most important, *the goal of an infinite game is to keep playing*. Think of the best parts of your life: your relationships, your creativity, your sexuality, your dreams, your adventures. The point is not to win, but to keep going. Winning is the death of the game.

Seen in light of Carse's thoughts, the whole expanse of our human trajectory presents a nonapocalyptic explanation for Fermi's troubling paradox. No doubt, many life-forms have destroyed themselves by "winning" finite games. But those that made it beyond the great filter may have done so by sensing the end before reaching it. They learned (or remembered) the central lesson of intelligent life before it was too late. A meal is as good as a feast. More is no better than enough. "Never satisfied" are not words to live by, but a colossal missing of the point that the game of life is not to be won. The point of life is the living of it. Keep playing, enjoy and prolong the experience. Maybe distant intelligences haven't been sending out signals because they realized that where they are, where they're from, is exactly where they want to be. There's no place like home, Toto. This response to the Fermi Paradox may also explain why human beings lived remarkably similar lives for 99 percent of our time on this planet. Life was good. Plenty birdies. Plenty fishies. Plenty mongongo nuts. No need to "advance" or "progress" from where we were. We were happy being there then.

We are at a crossroads, and going back is not an option. I envision three possible futures for *Homo sapiens*, the hominid that knows that it knows.

To one side lie Denial and Anger. Collapse: economic, ecological, political—a swirling, unstoppable category 7 hurricane of incompetence, indecency, greed, and devout ignorance masquerading as certainty. Maybe we're already so far down this path that there's nothing to be done but prepare for the coming storm. There are tipping points beyond which it doesn't matter whether we clean up our act. It's too late to get your shit together when it's already hit the fan. There are reports that methane that's been frozen since long before the dawn of civilization is already melting and bubbling to the surface of the oceans and rising in unstoppable vapors from Arctic permafrost. A large and growing community of scientists, environmentalists, and philosophers argue that we're already well into the terminal phase of civilization. Maybe right now is the moment of stunned, blinded silence after the lightning has flashed, but before the confirming thunder has clapped.

On the other side lie Bargaining and Depression: more of what got us here. We'll keep coming up with temporary fixes for the most immediate threats, and keep ignoring long-term trends as we have since our ancestors took their first steps out of the Garden onto the farm. As the destruction of the natural environment of this planet continues, we will evolve ever further from our organic origins, our fragile meat-bodies "upgraded" piece by piece with technological adaptations to a world increasingly toxic to living things. Today's titanium knees and hips will become tomorrow's implanted memory chips and subcutaneous GPS locators. The continued suffering of our animal souls will be increasingly numbed and medicated as the process proceeds to its inevitable conclusion. Tearful eyes will be replaced by unblinking electronic photodetectors that "see" far beyond the biological human visual spectrum, transmitting what they see to the all-knowing, all-seeing Orwellian swarm into which our descendants are absorbed so completely that individual human beings exist only in theory and prohibited memory. Again, we seem already to have taken steps far down this path.

Straight ahead lies Acceptance. What if we strategically bring hunter-gatherer thinking into our modern lives by, for example, replacing top-down corporate structures with peer progressive networks and horizontally organized collectives and building an infrastructure of nonpolluting locally generated energy? If *Homo sapiens sapiens* were to divert spending on weapons, redirecting resources into a global guaranteed basic income that incentivizes not having children, thus reducing global population intelligently and without coercion, we would be taking steps toward acceptance. Once we start down this road, every step would lead us closer to a future that recognizes, celebrates, honors, and replicates the origins and nature of our species. This is, as far as I can see, the only road home.

How likely is it that we will choose this path? Not very. But it's well within our capacities and budget to enact such programs, if sufficient shifts in consciousness demand it. If the notion of a step into the future being also a step into the past seems like a contradiction, consider that every winter day moves us both farther from and closer to the warmth of summer. The Enlightenment was simultaneously an extraordinarily progressive period *and* a celebration of the past embodied by ancient Rome and Greece. A movement to redesign the human zoo to reflect the origins and nature of *Homo sapiens* would represent a second, more brilliant Enlightenment, built to resonate with a more distant past.

"Everything the Power of the World does is done in a circle," said the great Lakota shaman Black Elk. "The sky is round, and I have heard that the earth is round like a ball, and so are the stars. The wind, in its greatest power, whirls. . . . The sun comes forth and goes down again in a circle. The moon does the same, and both are round. . . . The life of a man is a circle from childhood to childhood, and so it is in everything where power moves."

Acknowledgments

Books like this one are made of other books. So my first thanks go to the authors of those books—even (or especially) to those with whom I've disagreed in these pages. Our proposed answers may differ, but we share a passion for the deepest questions about the distant past and near future of our species.

It's strange that only one or two names appear on the cover of most books, in that they all result from the labor and attention of many people. The flaws and errors are easy enough to accomplish alone, but whatever is worthwhile here I owe to my friends, including my editor at Avid Reader Press, Ben Loehnen, whose vast patience I have stretched to its limits, and my agent, Andrew Stuart. Many of my friends and family have generously read (and often reread) drafts. I'd particularly like to thank Cacilda Jethá, Frank and Julie Ryan, Beth Ryan, Miguel Romero, Kyle Thiermann, John Stevens, Chris Bodenner, Anya Kaats, Erin Ginder-Shaw, Hunter Maats, Steve Herman, Celeste Phillips, Rick Moon, Steve Hellinger, Elena Arengo, Mary Smith, Simon Rex, Yeshe Perl, Don Mirra, Oliver Thorpe, Britt Winston, and Cheryl Hanna for their focused attention and tactfully brutal feedback. At one point, I ran out of gas on this project and Naomi Norwood rescued me. We spent many mornings shifting, restructuring, and refining. Without the benefit of her laser intellect and endless generosity, I doubt I'd have finished.

Notes and Further Reading

Introduction: Know Thy Species

4 Benjamin Franklin's thoughts on the seductive qualities of Indian life are from Walter Isaacson, *Benjamin Franklin: An American Life* (Simon & Schuster, 2003), p. 153.

4 Darwin's shock at seeing a Fuegian for the first time is from a letter to C. T. Whitley, July 23, 1834, http://www.darwinproject.ac.uk/letter/entry-250.

4 Button's given name was Orundellico, but the British called him "Jemmy Button" because he'd been bought from his uncle for the price of a single mother-of-pearl button. See *Savage*, by Nick Hazlewood (St. Martin's Press, 2000), for a gripping account of this man's incredible life, which apparently included leading the massacre of all aboard a missionary schooner thirty years later.

6 Forty-four percent of Americans earning between $40,000 and $100,000 per year told researchers that they couldn't come up with $400 in an emergency, and 27 percent of those making more than $100,000 said the same. Cited by Neal Gabler in *The Atlantic*, May 2016: "The Secret Shame of Middle-Class Americans," http://www.theatlantic.com/magazine/archive/2016/05/my-secret-shame/476415/.

6 Gould's denunciation of progress is from an essay called "On Replacing the Idea of Progress with an Operational Notion of Directionality," in M. H. Nitecki (ed.), *Evolutionary Progress* (University of Chicago Press, 1989).

7 Jared Diamond's line about industrial states not being necessarily better than hunter-gatherer tribes is from *Guns, Germs, and Steel* (W. W. Norton & Company, 1999), p. 18.

8 The apocalyptic article referred to is by Roy Scranton, "We're Doomed. Now What?," The Stone, *New York Times*, December 21, 2015, http://mobile.nytimes.com/blogs/opinionator/2015/12/21/were-doomed-now-what/?mc_cid=8fe1d86a0a&mc_eid=f97e8b93cc.

9 Ronald Wright's *A Short History of Progress* (Carroll & Graf, 2005) is a fantastic survey of how civilizations arise and wither away.

10 I came across the quote from Jonas Salk in John Durant's *The Paleo Manifesto* (Harmony, 2013), p. 28.

PART I: ORIGIN STORIES

15 "Our ancestors were not at one with nature. Nature tried to kill them and starve them out." This comes from an article called "Human Ancestors Were Nearly All Vegetarians," by Rob Dunn, published in *Scientific American* online, July 23, 2012, https://blogs.scientificamerican.com/guest-blog/human-ancestors-were-nearly-all-vegetarians/.

15 The dismal depiction of our prehistory, where everyone was apparently ugly, is from *Utopia for Realists*, by Rutger Bregman (Little, Brown and Company, 2017).

15 "Humanity is always moving forward": Will Martin, "This chart shows every major technological innovation in the last 150 years— and how they have changed the way we work," *Business Insider*, April 13, 2018. Left unaddressed is the fact that economic growth and quality of life don't necessarily increase in tandem. In the past century, for example, automation has arguably been the greatest driver of increases in productivity and economic growth, while also driving millions of people into poverty and despair.

Chapter 1: What We Talk About When We Talk About Prehistory

22 Jean Liedloff's thoughts on the "design" of the body expressing "expectations" are from her book *The Continuum Concept* (Da Capo Press, 1986), p. 23.

22 Daniel Everett's memoir of his time among the Pirahã is a fantastic read: *Don't Sleep, There Are Snakes: Life and Language in*

the Amazonian Jungle (Pantheon, 2008). For a brief introduction to Everett's work and the Pirahã, John Colapinto's article in *The New Yorker* is great: "The Interpreter: Has a Remote Amazonian Tribe Upended Our Understanding of Language?," April 9, 2007.

24 See *Limited Wants, Unlimited Means* (Island Press, 1997) for more on the economics of foraging societies. A collection of essays written primarily by anthropologists, and collected and edited by John Gowdy, an economist, the book offers an excellent overview of the behavioral and social characteristics common to foragers, and explains how these characteristics arise from a shared ecological context.

32 For more on miserable misers, see *The Paradox of Generosity*, by Christian Smith and Hilary Davidson (Oxford University Press, 2014).

33 Christopher Benfey's survey of utopian communities is "Building the American Dream," *New York Review of Books*, April 6, 2017.

Chapter 2: Civilization and Its Dissonance

41 For more on Nick Brooks and his work, see http://nickbrooks .org/.

41 You can read more of Jared Diamond's thoughts on the relative merits of civilization in "The Worst Mistake in the History of the Human Race," *Discover*, May 1999.

43 For more on the transition from foraging to agriculture, see Kirkpatrick Sale's *After Eden* (Duke University Press, 2006).

44 The quote from Eldredge is from Sale, pp. 97–98.

45 For more on Göbekli Tepe, see: "Paradise Regained?" *Fortean Times*, http://www.forteantimes.com/features/articles/449/gobekli_tepe _paradise_regained.html.

48 For more on the similarities between contemporary climate change and what was happening around thirteen thousand years ago, see G. W. K. Moore, K. Våge, R. S. Pickart, and I. A. Renfrew, "Decreasing Intensity of Open-Ocean Convection in the Greenland and Iceland Seas," *Nature Climate Change* 5 (2015): doi:10.1038/nclimate2688, published online June 29,

2015, http://www.nature.com/nclimate/journal/vaop/ncurrent/full/nclimate2688.html, and Thomas L. Delworth et al., "The Potential for Abrupt Change in the Atlantic Meridional Overturning Circulation," NOAA Geophysical Fluid Dynamics Laboratory, Princeton, New Jersey, https://www.gfdl.noaa.gov/bibliography/related_files/tdo802.pdf.

54 See Howard Zinn's excellent *People's History of the United States* (HarperCollins, 2003) for more on the first interactions between the Spanish and the Taíno.

55 For more on how the massive die-off among Native Americans may have triggered the Little Ice Age, see Alexander Koch, Chris Brierley, Mark M. Maslin, and Simon L. Lewis, "Earth System Impacts of the European Arrival and Great Dying in the Americas After 1492," *Quaternary Science Reviews* 207 (March 1, 2019): 13–36, https://doi.org/10.1016/j.quascirev.2018.12.004.

59 Margaret Ehrenberg on the status of female foragers: *Women in Prehistory* (University of Oklahoma Press, 1990), p. 65.

60 Charles Darwin, *The Descent of Man and Selection in Relation to Sex* (CreateSpace Independent Publishing Platform, 1871).

64 For more on the question of whether or not global economic inequality is improving, see Jason Hickel, "Is Global Inequality Getting Better or Worse? A Critique of the World Bank's Convergence Narrative," *Third World Quarterly* (2017, http://dx.doi.org/10.1080/01436597.2017.1333414). Also by Hickel, "Exposing the Great 'Poverty Reduction' Lie," Al Jazeera, August 21, 2014, https://www.aljazeera.com/indepth/opinion/2014/08/exposing-great-poverty-reductio-201481211590729809.html. A book-length discussion of the issue is *The Growth Delusion: The Wealth and Well-Being of Nations* by David Pilling (Tim Duggan Books, 2018).

64 For more on the situation with Coca-Cola in India, see https://www.theguardian.com/environment/2014/jun/18/indian-officials-coca-cola-plant-water-mehdiganj. A few excellent, nuanced explorations of the situation in Cochabamba can be found here: http://www.newyorker.com/magazine/2002/04/08/leasing-the-rain, and here: http://www.ucpress.edu/content/chapters/11049.ch01.pdf, and in this *Frontline* documentary: http://www.pbs.org/frontlineworld/stories/bolivia/thestory.html.

66 Despite his shaky calculations, Malthus was correct in the most basic and chilling sense: Population grows; the planet doesn't.

68 Mark Lilla's description of Hobbes's life and times is from *The Stillborn God: Religion, Politics, and the Modern West* (Knopf, 2007).

70 Richard Dawkins's freak-out about rasping parasites and so on can be found in *River Out of Eden* (Basic Books, 2008), pp. 131–32.

74 See Eric Michael Johnson's article for more on the interplay of selfishness and group welfare: http://evonomics.com/ayn-rand-vs -anthropology/#comment-2389720011.

81 Kelly's explanation of "egalitarianism" is from *The Foraging Spectrum: Diversity in Hunter-Gatherer Lifeways* (Washington, D.C.: Smithsonian Institution Press, 1995), p. 296.

82 Frans de Waal has been studying the primate impulse toward justice for several decades. See, for example, *Chimpanzee Politics: Power and Sex among Apes* (Johns Hopkins University Press, 2007) and *The Bonobo and the Atheist: In Search of Humanism Among the Primates* (W. W. Norton & Company, 2013).

82 The story about "cooling the heart" of a boastful hunter is from Richard Lee (*The !Kung San: Men, Women, and Work in a Foraging Society*, 1979, pp. 244–46), cited in *Hierarchy in the Forest*, by Christopher Boehm (Harvard University Press, 1999), p. 45. (Previous block quote, same source.)

82 The egalitarianism of the Inuit is from Kent Flannery and Joyce Marcus, *The Creation of Inequality* (Harvard University Press, 2012), p. 24.

83 Boehm hammers his point about the generosity of leaders in a paragraph that offers a sense of how universally selfless qualities are admired in band-level societies around the world and how much scholarship the neo-Hobbesians have to ignore to stick to their "selfish infiltrator" theory:

> Among the Coeur d'Alene, wisdom, *generosity*, and honesty were valued (Teit 1930:152–153). A Mescalero Apache chief was good at talking and thinking, *generous*, and respectful (Basehart 1970:99), while Godwin says of the Apache that a chief should be capable as a warrior and hunter and successful economically, but also *generous*, impartial, patient, and in control of his temper (see Basso 1971:14). Denig (1930:449)

says of the Assiniboin that *parsimony*, along with exceptional meanness, was criticized—and in fact the chief tended to be the poorest man in the camp. Arapaho leaders were expected to be brave, trustworthy, *willing to share food unselfishly*, and to have good sense and judgment (Hilger 1952:190). Jenness (1935:2) delineates how an Ojibwa chief would *provide for a needy family from his own resources*, or arrange contributions from other band members. For the Australian Pintupi, Myers (1980) makes the case that a primary role of chiefs and elders was to *take care of* other Aborigines. For the Kalahari !Kung, Marshall (1967:38) says that headmanship is not much desired and that the leader has to be *generous* and careful not to stand out. (All emphases in the original.)

83 Peter Bogucki, *The Origins of Human Society* (Wiley-Blackwell, 2000), p. 77.

84 Sebastian Lippold et al., "Human Paternal and Maternal Demographic Histories: Insights from High-Resolution Y Chromosome and mtDNA Sequences," *Investigative Genetics* 5, no. 13 (2014), doi:10.1186/2041-2223-5-13. The electronic version of this article is the complete one and can be found online at http://www.investigativegenetics.com/content/5/1/13.

84 The authors of a 2008 review (Aureli et al., "Fission-Fusion Dynamics: New Research Frameworks," *Current Anthropology* 49, no. 4 [2008]: 628) summarize the anthropological literature like this:

> Fission-fusion dynamics are . . . typical of modern humans, including hunter-gatherers (Marlowe 2005), although they are not often explicitly recognized. The following quote captures this apparent anomaly: "Fission-fusion sociality seems so natural and necessary to humans—including anthropologists—that it hardly demands explanation, if it is noticed at all." (Rodseth et al. 1991, 238). The sharing of this flexible social nature with our closest living relatives suggests that fission-fusion dynamics were characteristic of the social system of the last common ancestor of chimpanzees, bonobos, and modern humans.

84 The Nurit Bird-David quote is from *Limited Wants, Unlimited Means*, p. 130.

86 Crockett's research: "Most People Would Rather Harm Themselves Than Others for Profit," UCL News, November 18, 2014, http://www.ucl.ac.uk/news/news-articles/1114/181114-rather-harm-selves-than-others-for-profit#sthash.KBwLtz4x.dpuf.

87 Cited in *Tribe*, by Sebastian Junger (Twelve, 2016).

88 Frans B. M. de Waal, "Morality and the Social Instincts: Continuity with the Other Primates," Tanner Lectures on Human Values, delivered at Princeton University, November 19–20, 2003.

Chapter 3: The Myth of the Savage Savage (Declaring War on Peace)

91 Parts of this section are adapted from an essay I wrote that originally appeared online as "Hobbled by Hobbes: How Chimpanzees Became Nasty, Brutish and Short," published by the Evolution Institute, https://evolution-institute.org/hobbled-by-hobbes-how-chimpanzees-became-nasty-brutish-and-short/.

94 Similarities between bonobos and humans include having sex face-to-face, kissing, mothers passing infants to other females soon after birth, frequent same-sex interactions, etc. For much more on our shared traits with bonobos, see Frans de Waal and Frans Lanting, *Bonobo: The Forgotten Ape* (University of California Press, 1997).

94 In *Untrue* (Little, Brown, Spark, 2018), Wednesday Martin reports observations by primatologist Amy Parish that may be interpreted as coercive sexual interactions initiated by female bonobos with unwilling males. So the closest thing to "rape" in this species may be overly persistent females harassing males.

95 Sapolsky's description of the peaceful baboon troop can be found in many places, including this article he wrote for *Yes* magazine: https://www.yesmagazine.org/issues/can-animals-save-us/warrior-baboons-give-peace-a-chance.

96 Douglas P. Fry and Patrik Söderberg's article is "Myths about Hunter-Gatherers Redux: Nomadic Forager War and Peace," *Journal of Aggression, Conflict and Peace Research* 6, no. 4 (2014): 255–66.

98 Steven Pinker has the unfortunate habit of mislabeling horticulturalists—with their gardens, domesticated animals, and villages—as hunter-gatherers, who have none of these things. This

mislabeling is extremely problematic in that accumulated wealth is worth fighting over. See R. Brian Ferguson's essay "Pinker's List: Exaggerating Prehistoric War Mortality" in *War, Peace, and Human Nature*, edited by Douglas Fry (Oxford University Press, 2013), pages 112–31, for more on the details and consequences of this confusion.

98 The article I refer to by Bowles is "Did Warfare Among Ancestral Hunter-Gatherers Affect the Evolution of Human Social Behaviors?" *Science* 324 (2009): 1293–98.

100 For more on rates of same-species lethality among mammals, see José María Gómez, Miguel Verdú, Adela González-Megías, and Marcos Méndez, "The Phylogenetic Roots of Human Lethal Violence," *Nature* 538 (October 13, 2016): 233–37, https://www.nature.com/articles/nature19758.

Chapter 4: The Irrational Optimist

102 For scholarship on the seemingly inevitable collapse of civilizations, see, for example, Joseph Tainter's *The Collapse of Complex Societies* (Cambridge University Press, 1990).

105 Ridley claims the air and water are cleaner now than ten thousand years ago, yet *The Lancet* published a robust research report showing that pollution is the world's leading environmental cause of disease, causing 9 million premature deaths in 2015, and 16 percent of all deaths worldwide. This is three times the toll taken by AIDS, tuberculosis, and malaria combined, and fifteen times more than the toll of all wars and other forms of violence. Most of these deaths were in low- and middle-income countries, and in the poor communities of rich countries, which don't seem to be included in Ridley's calculations. https://www.thelancet.com/journals/lancet/article/PIIS0140-6736(17)32345-0/fulltext.

106 For more on the health of foragers, see, for example, P. Carrera-Bastos et al., "The Western Diet and Lifestyle and Diseases of Civilization," *Research Reports in Clinical Cardiology* 2 (2011): 15–35. Another excellent source is *Health and the Rise of Civilization*, by Mark Nathan Cohen (Yale University Press, 1989).

106 Bodley's research is presented in *Victims of Progress* (Rowman & Littlefield, 2014).

108 For more on tooth decay, see K. Gruber, "Oral Mystery: Are Agriculture and Rats Responsible for Tooth Decay?" *Scientific American*, February 6, 2013; D. L. Greene, G. H. Ewing, and G. J. Armelagos, "Dentition of a Mesolithic Population from Wadi Halfa, Sudan," *American Journal of Physical Anthropology* 27 (1967): 41–55; and W. Price, *Nutrition and Physical Degeneration* (Price-Pottenger Nutrition, 2008).

Also, the recent explosion in myopia—up 66 percent in the United States in the past thirty years—appears to be due to our modern appetite for indoor screens rather than outdoor sunlight. So the claim that eyeglasses are a good reason to prefer modern life falls into the same nonsense category as dental care.

109 As Frank Marlowe explains in *The Hadza: Hunter-Gatherers of Tanzania* (University of California Press, 2010), Hadza women reach puberty around eighteen, bear an average of 6.2 children (plus two to three noticeable miscarriages) starting at nineteen, and hit menopause in their early forties. Babies typically breast-feed for about four years. So of these twenty-five years of reproductive life, roughly twenty are spent lactating and 4.5 pregnant, resulting in fewer than a dozen menstruations in a woman's lifetime. Other studies have estimated around a hundred menstruations for foragers. For example, a study of the Dogon of Mali based on 57 women estimated the median number of lifetime menses at 109. Beverly I. Strassman, "The Biology of Menstruation in Homo Sapiens: Total Lifetime Menses, Fecundity, and Nonsynchrony in a Natural-Fertility Population," *Current Anthropology* 38, no. 1 (February 1997): 123–29.

109 To be clear, my discussion of how increased menstrual cycles may affect cancer rates is not meant as a critique of hormonal contraception or to advocate early pregnancy, but merely to show ways in which modern advances can have unexpected consequences. For much more on these unintended consequences, see Daniel Lieberman's *The Story of the Human Body* (Pantheon, 2013).

109 The information about health of the Waorani is from J. W. Larrick, J. A. Yost, J. Kaplan, G. King, and J. Mayhall, "Patterns of Health and Disease Among the Waorani Indians of Eastern Ecuador," *Medical Anthropology* 3, no. 2 (May 12, 2010): 147–89.

Also see: http://www.nytimes.com/1983/11/08/science/a
-doctor-in-the-amazon-probes-for-genetic-links-to-disease
.html. (Despite the amazing general health of the Waorani, the
scientists reported that they seemed to lack an enzyme that
protects teeth, so their oral health wasn't so good.)

110 The variety of the !Kung diet is discussed by Jared Diamond in
The Third Chimpanzee (HarperCollins, 1992), p. 166.

110 The data on world hunger are from: http://www.worldhunger
.org/articles/Learn/world%20hunger%20facts%202002.htm.

111 For more on how a little hunger can be a good thing, see Krista
A. Varady and Marc K. Hellerstein, "Alternate-Day Fasting and
Chronic Disease Prevention: A Review of Human and Animal
Trials," *American Journal of Clinical Nutrition* 86, no. 1 (July
2007): 7–13, http://ajcn.nutrition.org/content/86/1/7.full. This
article includes detailed references for each of the specific benefits
of calorie restriction.

113 For more on child mortality among the Hadza, see Marlowe,
Hadza, p. 150.

114 The definitive study of longevity among foragers is from
Michael Gurven and Hillard Kaplan, "Longevity Among
Hunter-Gatherers: A Cross-Cultural Examination," *Population
and Development Review*, May 29, 2007, http://onlinelibrary
.wiley.com/doi/10.1111/j.1728-4457.2007.00171.x/abstract.

114 The longevity study looking at anatomical similarities among
primates is: James R. Carey, "Life Span: A Conceptual Overview,"
in *Life Span: Evolutionary, Ecological, and Demographic Perspectives*, edited by James R. Carey and Shripad Tuljapurkar (Population Council, 2003). Available online at https://pingpdf.com/pdf
-life-span-evolutionary-ecological-and-population-council.html.

116 The data on foundling hospitals are from Sandra Newman,
"Infanticide," *Aeon*, November 27, 2017, https://aeon.co/essays
/the-roots-of-infanticide-run-deep-and-begin-with-poverty.

117 Abortion in China reported by *China Daily* and cited in Vicky
Jiang, "Of the 13 Million Abortions in China, Most Are Forced,"
Epoch Times, December 9, 2012, http://www.theepochtimes.com
/n2/china-news/one-child-policy-abortions-in-china-most-are
-forced-21819-all.html.

117 Hrdy quoted by Eric Michael Johnson in "Raising Darwin's Consciousness: Sarah Blaffer Hrdy on the Evolutionary Lessons of
Motherhood," *Scientific American*, March 16, 2012, http://blogs
.scientificamerican.com/primate-diaries/2012/03/16/raising
-darwins-consciousness-sarah-blaffer-hrdy-on-the-evolutionary
-lessons-of-motherhood/.

PART III: REFLECTIONS IN AN ANCIENT MIRROR (BEING HUMAN)

122 David Dobbs, "Die, Selfish Gene, Die," *Aeon*, December 3, 2013,
http://aeon.co/magazine/science/why-its-time-to-lay-the-selfish
-gene-to-rest/.
123 S. Zuckerman, *The Social Life of Monkeys and Apes* (Mellon
Press, 2011).
124 Eric Michael Johnson, "Frans de Waal on Political Apes, Science
Communication, and Building a Cooperative Society,"
Scientific American, July 11, 2011, http://blogs.scientific
american.com/primate-diaries/httpblogsscientificamerican
comprimate-diaries20110711frans-de-waal/.
125 Beth Mole, "'Is Curing Patients a Sustainable Business Model?'
Goldman Sachs Analysts Ask," *Ars Technica*, April 12, 2018,
https://arstechnica.com/tech-policy/2018/04/curing-disease
-not-a-sustainable-business-model-goldman-sachs-analysts
-say/?comments=1&post=35150219.

Chapter 5: The Naturalistic Fallacy Fallacy

129 The business consultants explaining how cultivating dissatisfac-
tion is good business are quoted in Stuart Ewen's *Captains of
Consciousness: Advertising and the Social Roots of the Consumer
Culture* (McGraw-Hill, 1976), p. 39.

Chapter 6: Born to Be Wild

134 The information about Efé adult contact with infants is from
http://anthro.vancouver.wsu.edu/media/Course_files/anth-302
-barry-hewlett/melkonner.pdf.

Chapter 7: Raising Hell

143 Richard Schweid, in his 2016 book *Invisible Nation: Homeless Families in America*, reports that 2.5 million children experience homelessness every year in the United States, sleeping with their families in cars, motel rooms, or packed into the home of whatever relative will take them in. Study after study shows that homelessness is both mentally and physically unhealthy for children and that the "toxic stress" of homelessness can have deleterious effects on them even after they grow into adults. While it's tempting to see the disregard for American children as an unfortunate coincidence, there are only two nations in the world that steadfastly refuse to ratify the UN Convention on the Rights of the Child: South Sudan and the United States. While South Sudan can point to a lack of funding to implement even the most basic protections for children, the United States has no such excuse. In the six years from 2009 to 2015, both America's wealth *and* its population of homeless children grew by roughly 60 percent. Study after study has demonstrated that wealth disparity is correlated with infanticide. The United States, often described as the world's wealthiest nation, leads the developed world with a maternal infanticide rate of eight deaths for every hundred thousand live births—twice Canada's rate. Again, this is not merely a result of poverty. The highest rates of maternal infanticide are found not in the poorest states, but in those with the widest disparities in wealth. Babies born to impoverished women in Colorado, Oklahoma, and New York, for example, are three to five times more likely to be killed by their mothers—as compared to the national average.

146 The data on overtreatment of ADHD are from Ryan D'Agostino's heartbreaking essay called "The Drugging of the American Boy," *Esquire*, March 27, 2014.

Chapter 8: Turbulent Teens

153 The accounts of Kellogg's child abuse are from John Money's *The Destroying Angel: Sex, Fitness & Food in the Legacy of Degeneracy*

Theory, Graham Crackers, Kellogg's Corn Flakes & American Health History (Prometheus Books, 1985).

155 Stephen T. Asma, "The Weaponized Loser," *Aeon*, June 27, 2016, https://aeon.co/essays/humiliation-and-rage-how-toxic -masculinity-fuels-mass-shootings.

156 Alek Minassian, the man who killed ten people by driving a van down a busy sidewalk in Toronto in 2018, gave Elliot Roger a shout-out, calling attention to the so-called Incel movement, referring to the involuntary celibacy that they shared.

156 Jane Lewis and Trudie Knijn's research on sex ed in the Netherlands, England, and Wales can be found in the *Oxford Review of Education* 29, no. 1 (2003): 113–50, or online at https://doi .org/10.1080/03054980307431. Amy T. Schalet's book *Not Under My Roof: Parents, Teens, and the Culture of Sex* (University of Chicago Press, 2011) is another source of excellent insight, comparing sex ed in the Netherlands and the United States. Some of the data concerning adolescents comes from *Adolescence: An Anthropological Inquiry* (Free Press, 1991), by Alice Schlegel and Herbert Barry III.

158 Whitlock is quoted in http://time.com/4547322/american-teens -anxious-depressed-overwhelmed/.

Chapter 9: Anxious Adults

161 Jonnie Hughes talks about his experiences with the *Insect Tribe* in *On the Origin of Tepees: The Evolution of Ideas (and Ourselves)* (Free Press, 2011).

163 See *Limited Wants, Unlimited Means* for more on how the behavior of foragers is a counterexample of what mainstream economic theory predicts.

165 Martin Gusinde's quote is from *Stone Age Economics*, by Marshall Sahlins (revised edition, Routledge, 2013).

175 The M&M's study is reported in James Surowiecki's "Downsizing Supersize," *New Yorker*, August 13, 2012, http://www.newyorker .com/magazine/2012/08/13/downsizing-supersize.

176 The camera study is reported in Itamar Simonson and Amos Tversky, "Choice in Context: Tradeoff Contrast and Extreme-

ness Aversion," *Journal of Marketing Research* 29, no. 3 (1992): 281–95.

176 Gary Rivlin's article is "In Silicon Valley, Millionaires Who Don't Feel Rich," *New York Times*, August 5, 2007. http://www.nytimes .com/2007/08/05/technology/05rich.html?pagewanted=all.

181 For more on the corrosive effects of economic inequality, see Stéphane Côté, Julian House, and Robb Willer, "High Economic Inequality Leads Higher-Income Individuals to Be Less Generous," *PNAS*, November 23, 2015, http://www.pnas.org/content /early/2015/11/18/1511536112; J. Moll et al., "Human Fronto-Mesolimbic Networks Guide Decisions About Charitable Donation," *PNAS* 103, no. 42 (October 17, 2006): 15623–28; J. G. Miller, S. Kahle, and P. D. Hastings, "Roots and Benefits of Costly Giving: Children Who Are More Altruistic Have Greater Autonomic Flexibility and Less Family Wealth," *Psychological Science* 26, no. 7 (July 2015): doi: 10.1177/0956797615578476; Shankar Vedantam, "If It Feels Good to Be Good, It Might Be Only Natural," *Washington Post*, May 28, 2007, http://www.washingtonpost.com/wp-dyn /content/article/2007/05/27/AR2007052701056.html; and Richard Wilkinson and Kate Pickett, *The Spirit Level: Why Greater Equality Makes Societies Stronger* (Bloomsbury Press, 1999).

184 For more on the primate origins of morality, see Frans de Waal, *The Bonobo and the Atheist* (W. W. Norton & Company, 2013).

186 Michael Lewis's essay "Extreme Wealth Is Bad for Everyone—Especially the Wealthy," *New Republic*, November 12, 2014, can be found online at http://www.newrepublic.com/article/120092/ billionaires-book-review-money-cant-buy-happiness.

187 Quotes from Chris Benderev, "When Power Goes to Your Head, It May Shut Out Your Heart," NPR, https://www.npr.org/2013/08/10/ 210686255/a-sense-of-power-can-do-a-number-on-your-brain.

187 For more on the deleterious effects of social isolation, see https:// www.campaigntoendloneliness.org/threat-to-health/.

189 For more on Piff's research, see his TED Talk: http://www .ted.com/talks/paul_piff_does_money_make_you_mean ?language=en.

190 Robb Willer is quoted in http://www.berkeley.edu/news/media /releases/2009/12/08_survival_of_kindest.shtml.

Chapter 10: All's Well That Ends Well

194 Ernest Becker, *The Denial of Death* (Free Press, 1997).

194 You can hear my conversation with Sheldon Solomon in episode 154 of *Tangentially Speaking*, https://chrisryanphd .com/tangentially-speaking/2015/11/22/154-sheldon-solomon -terror-management-theory.

195 You can listen to my conversation with Caitlin Doughty in *Tangentially Speaking*, episode 90: https://chrisryanphd.com /tangentially-speaking/2014/9/16/90-caitlin-doughty-smoke -gets-in-your-eyes.

195 My discussion of end-of-life care draws from A. E. Singer, D. Meeker, J. M. Teno, J. Lynn, J. R. Lunney, and K. A. Lorenz, "Symptom Trends in the Last Year of Life from 1998 to 2010: A Cohort Study," *Annals of Internal Medicine* 162, no. 3 (February 3, 2015): 175–83, doi: 10.7326/M13-1609; Jason Millman, "It's Only Getting Worse to Die in America," *Washington Post*, February 3, 2015, http://www.washingtonpost.com/news /wonkblog/wp/2015/02/03/its-only-getting-worse-to-die-in -america/; Amber E. Barnato et al., "Trends in Inpatient Treatment Intensity among Medicare Beneficiaries at the End of Life," *Health Services Research* 39, no. 2 (April 2004): 363–76; Craig Brown, "Our Unrealistic Views of Death, Through a Doctor's Eyes," *Washington Post*, February 17, 2012, https://www .washingtonpost.com/opinions/our-unrealistic-views-of-death -through-a-doctors-eyes/2012/01/31/gIQAeaHpJR_story.html? noredirect=on&utm_term=.e2d372214371; and J. J. Gallo, J. B. Straton, M. J. Klag, L. A. Meoni, D. P. Sulmasy, N. Y. Wang, and D. E. Ford, "Life-Sustaining Treatments: What Do Physicians Want and Do They Express Their Wishes to Others?" *Journal of the American Geriatrics Society* 51, no. 7 (July 2003): 961–69.

A chart of doctor responses from the Precursors Study (http:// pages.jh.edu/~jhumag/0601web/study.html); Ken Murray, "How Doctors Die," *Zocal*, November 30, 2011, http://www.zocalo publicsquare.org/2011/11/30/how-doctors-die/ideas/nexus/; Jennifer S. Temel et al., "Early Palliative Care for Patients with Metastatic Non–Small-Cell Lung Cancer," *New England Journal of Medicine*

363 (August 19, 2010): 733–42, doi:10.1056/NEJMoa1000678; Nina Bernstein, "Fighting to Honor a Father's Last Wish: To Die at Home," *New York Times*, September 25, 2014, https://www .nytimes.com/2014/09/26/nyregion/family-fights-health-care -system-for-simple-request-to-die-at-home.html; "The Long Goodbye," an interview by Sam Mowe in the *Sun Magazine*, no. 460, April 2014; Baohui Zhang et al., "Health Care Costs in the Last Week of Life: Associations with End-of-Life Conversations," *Archives of Internal Medicine* 169, no. 5 (March 9, 2009): 480–88; and Marina Gafanovich, "End-of-Life Care Constitutes Third Rail of U.S. Health Care Policy Debate," Medicare NewsGroup, September 17, 2015, available at http://www.mynycdoctor.com/end-of-life-care -constitutes-third-rail-of-us-health-care-policy-debate.

201 For more on the Gilbert case, see https://people.com/archive/ the-agony-did-not-end-for-roswell-gilbert-who-killed-his-wife -to-give-her-peace-vol-27-no-2/.

202 Ezekiel J. Emanuel's essay "Why I Hope to Die at 75," *Atlantic*, October 2014, is online here: http://www.theatlantic.com/magazine /archive/2014/10/why-i-hope-to-die-at-75/379329/.

203 Eileen Crimmins's research quoted in "Lifespan and Healthspan: Past, Present, and Promise," from *Gerontologist* 55, no. 6 (December 2015): 901–11.

Chapter II: In the Absence of the Sacred

205 The chapter title is a lovely phrase from the title of a book by Jerry Mander (Random House, 1991).

207 Aldous Huxley's quote is from *Brave New World* (Harper, 2017).

208 My conversation with Luhrmann is episode 114 of *Tangentially Speaking*, https://chrisryanphd.com/tangentially-speaking/2015 /3/1/114-dr-tanya-luhrmann-psychological-anthropologist.

212 If you'd like to read more about the connections between schizophrenia and shamanism, take a look at *The Shaman's Doorway: Opening Imagination to Power and Myth*, by Stephen Larsen (Inner Traditions, 1998), and Roger Walsh's *The World of Shamanism: New Views of an Ancient Tradition* (Llewellyn Publications, 2007).

217 For more on Timothy Tyler's case, see https://www.psymposia
.com/magazine/timothy-tyler-clemency-obama-lsd/.

217 For more on the Bob Riley case, the court transcript is here:
http://faculty.rwu.edu/dzlotnick/profiles/longstaff_files/longstaff
.html#_ftn1.

218 For more on the struggles to win approval of MDMA, see Jac-
queline Ronson, "When Psychiatrists Fought Like Hell to Keep
MDMA Legal," *Inverse*, August 6, 2016, https://www.inverse.com
/article/19195-mdma-1980s-court-battle-psychiatrists-versus-dea.

219 You can hear my conversation with Rick Doblin in episode 98 of
Tangentially Speaking, https://chrisryanphd.com/tangentially
-speaking/2014/11/3/98-rick-doblin-maps.

221 Dan Baum, "Legalize It All: How to Win the War on Drugs," *Harper's Magazine*, April 2016, https://harpers.org/archive/2016/04
/legalize-it-all/.

224 See Michael Pollan's review of the MDMA and psilocybin research
in *The New Yorker*, https://www.newyorker.com/magazine
/2015/02/09/trip-treatment. Several of the quotations in this
section are from Pollan's article. Also see his subsequent book on
the potential beneficial uses of psychedelics: *How to Change Your
Mind: What the New Science of Psychedelics Teaches Us About
Consciousness, Dying, Addiction, Depression, and Transcendence*
(Penguin, 2018).

226 Charles Johnston spoke to me on *Tangentially Speaking*, episode
175, https://chrisryanphd.com/tangentially-speaking/2016/3/28
/charles-johnston-former-heroin-addict.

227 Quoted in Ronald Piana, "Researchers Discuss Pilot Study on
Hallucinogenic Therapies for Cancer Anxiety," *ASCO Post* 6, no. 8
(May 10, 2015), http://www.ascopost.com/issues/may-10,-2015
/researchers-discuss-pilot-study-on-hallucinogenic-therapies
-for-cancer-anxiety.aspx.

229 For more on Summergrad's embrace of psychedelics, see https://
maps.org/news/media/6313-high-times-the-mainstreaming
-of-psychedelics or Michael Pollan's *How to Change Your Mind*
(Penguin, 2018).

229 For more on how the gods of foragers differ from the God of
farmers, see this paper published in *Nature* in 2019 by Harvey

Whitehouse and colleagues: "Complex Societies Precede Moralizing Gods Throughout World History," http://doi.org/10.1038/s41586-019-1043-4.

234 The anecdote about the "God particle" is from Mark Memmott, "Here We Go Again: Has Misnamed 'God Particle' Finally Been Found?," NPR, December 12, 2011, http://www.npr.org/sections/thetwo-way/2011/12/12/143571097/here-we-go-again-has-misnamed-god-particle-finally-been-found.

234 T. V. Rajan's "The Myth of Mechanism" is from *The Scientist*, June 2001, http://www.the-scientist.com/?articles.view/articleNo/13463/title/The-Myth-of-Mechanism/.

234 Farber's quote is cited in Emily Eakin, "Bacteria on the Brain," *The New Yorker*, December 7, 2015, http://www.newyorker.com/magazine/2015/12/07/bacteria-on-the-brain.

236 The data on religiosity and health are from Jeffrey S. Levin, "How Religion Influences Morbidity and Health: Reflections on Natural History, Salutogenesis and Host Resistence," http://deathandreligion.plamienok.sk/files/69-HOW%20RELIGION%20INFLUENCES%20MORBIDITY%20AND%20HEALTH.pdf.

238 The Kickstarter information is from https://www.kickstarter.com/help/stats.

239 For more on emergent phenomena, see Kevin Kelly, "Out of Control: The New Biology of Machines, Social Systems and the Economic World," https://kk.org/mt-files/books-mt/ooc-mf.pdf, p. 15.

Conclusion: A Necessary Utopia

241 Isaac Asimov's quote is from *A World of Ideas* on PBS, interviewed by Bill Moyers, May 26, 1989.

246 The story about the kids who were stranded on the island is from S. Agnelli, *Street Children: A Growing Urban Tragedy: A Report for the Independent Commission on International Humanitarian Issues* (Weidenfeld & Nicholson, 1986).

250 The news about potentially unstoppable methane eruptions is explained in "Horrific Methane Eruptions in East Siberian Sea," *Arctic News*, August 13, 2014, http://arctic-news.blogspot.ie/2014/08/horrific-methane-eruptions-in-east-siberian-sea.html.

Index

Page references in italics indicate illustrations.

About the Author

Christopher Ryan, PhD, and his work have been featured just about everywhere, including MSNBC, Fox News, CNN, NPR, HBO, Netflix, the *New York Times*, the *Times of London*, *Playboy*, the *Washington Post*, *Time*, *Newsweek*, the *Atlantic*, *Salon*, TED, and *Big Think*. He is coauthor of the *New York Times* bestseller *Sex at Dawn* (translated into twenty-two languages) and hosts a weekly podcast, *Tangentially Speaking*, featuring conversations with people ranging from famous comics to bank robbers to drug smugglers to porn stars to authors to astrophysicists.